MAIZE GERMPLASM – CHARACTERIZATION AND GENETIC APPROACHES FOR CROP IMPROVEMENT

Edited by **Mohamed El-Esawi**

Maize Germplasm – Characterization and Genetic Approaches for Crop Improvement

http://dx.doi.org/10.5772/intechopen.68373
Edited by Mohamed El-Esawi

Contributors

Raul Rodriguez-Herrera, Mariela R. Michel, Marisol Cruz-Requena, Marcelino Avendaño-Sanchez, Victor González-Vazquez, Adriana Carolina Flores Gallegos, Cristobal Aguilar, Jose Espinoza-Velázquez, Meimei Lin, Eastonce T Gwata, Abera Wende, Hussein Shimelis, Wilian Henrique Diniz Buso, Luciana Borges E Silva, Asima Gazal, Zahoor Dar, Ajaz A. Lone, Mohamed A. El-Esawi

Notice

Statements and opinions expressed in the chapters are these of the individual contributors and not necessarily those of the editors or publisher. No responsibility is accepted for the accuracy of information contained in the published chapters. The publisher assumes no responsibility for any damage or injury to persons or property arising out of the use of any materials, instructions, methods or ideas contained in the book.

First published in London, United Kingdom, 2018 by IntechOpen
IntechOpen is the global imprint of INTECHOPEN LIMITED, registered in England and Wales, registration number: 11086078, The Shard, 25th floor, 32 London Bridge Street
London, SE19SG – United Kingdom
Printed in Croatia

British Library Cataloguing-in-Publication Data
A catalogue record for this book is available from the British Library

Additional hard copies can be obtained from orders@intechopen.com

Maize Germplasm – Characterization and Genetic Approaches for Crop Improvement, Edited by Mohamed El-Esawi
p. cm.
Print ISBN 978-1-78923-038-3
Online ISBN 978-1-78923-039-0

For EU product safety concerns:
IN TECH d.o.o., Prolaz Marije Krucifikse Kozulić 3, 51000 Rijeka, Croatia,
info@intechopen.com or visit our website at intechopen.com.

Meet the editor

Dr. Mohamed Ahmed El-Esawi is currently a visiting research fellow at the University of Cambridge in the United Kingdom and an assistant professor of Molecular Biology and Genetics at Botany Department of Tanta University in Egypt. Dr. El-Esawi received his BSc and MSc degrees from Tanta University and his PhD degree in Plant Genetics and Molecular Biology from Dublin Institute of Technology in Ireland. Afterward, Dr. El-Esawi joined the University of Warwick in the United Kingdom, University of Sorbonne (Paris VI) in France, and University of Leuven (KU Leuven) in Belgium as a visiting research fellow. His research focuses on plant genetics, genomics, molecular biology, molecular physiology, developmental biology, plant-microbe interaction, and bioinformatics. He has authored several international journal articles and book chapters and participated in more than 60 conferences and workshops worldwide. Dr. El-Esawi is currently involved in several research projects on biological sciences.

Contents

Preface

Maize is an important staple food crop worldwide. It is the third most important cereal crop after wheat and rice and is economically used for both livestock feeds and human consumption. The latest maize research has opened up new opportunities for crop improvement. Oilseed crops, including maize, also play an important role in the agricultural economy. Globally, the demand for vegetable oils is increasing due to the increasing per capita consumption of oil in our daily diets and its use as biofuels. A range of new technologies have been developed to enhance the productivity of this crop.

This book brings together recent works and advances that have recently been made in the dynamic fields of genetic characterization, molecular breeding, genetic engineering technologies, and mapping of agronomic traits of global maize germplasm. It also provides new insights into and sheds new light regarding the current research trends and future research directions in maize. This book will provoke interest in many readers, researchers, and scientists, who can find this information useful for the advancement of their research works toward maize improvement.

The book includes six chapters. The first chapter "Introduction to Biotechnological Approaches for Maize Improvement" presents an introduction to the genetic and biotechnological approaches developed to enhance maize productivity. The second chapter "Polyembryony in Maize: A Complex, Elusive, and Potentially Agronomical Useful Trait" provides a review of the present literature on polyembryony phenomenon and discusses its applications and possible causes in maize. The third chapter "Molecular Breeding for Abiotic Stresses in Maize (*Zea mays* L.)" discusses developments in molecular breeding technologies for developing and improving abiotic stress resilience in maize. The fourth chapter "Genetic Variability for Resistance to Leaf Blight and Diversity among Selected Maize Inbred Lines" studies the genetic variability in reaction to Turcicum leaf blight among maize inbred lines under field conditions as well as evaluates the diversity of selected medium to late maturity tropical maize inbred lines for hybrid breeding using selected SSR markers. The fifth chapter "Use of Technology to Increase the Productivity of Corn in Brazil" gives a comprehensive overview on the technologies developed to increase the corn productivity in Brazil. The sixth chapter "Impacts of Nitrogen Fertilization and Conservation Tillage on the Agricultural Soils of the United States" discusses the effects of N fertilization and conversion of management practice from conventional tillage (CT) to no till (NT) on soil organic carbon stocks in the United States.

The book editor would like to thank Ms. Martina Usljebrka, Publishing Process Manager, for her wholehearted cooperation in the publication of this book.

Mohamed Ahmed El-Esawi, PhD
Sainsbury Laboratory, University of Cambridge, Cambridge, United Kingdom
Botany Department, Faculty of Science, Tanta University, Egypt

Chapter 1

Introductory Chapter: Introduction to Biotechnological Approaches for Maize Improvement

Mohamed A. El-Esawi

Additional information is available at the end of the chapter

http://dx.doi.org/10.5772/intechopen.74871

1. Introduction

Maize (*Zea mays* L.) is an important staple food crop worldwide. It is the third most important cereal crop after wheat and rice [1]. Maize is economically used for both livestock feeds and human consumption. The agricultural production of maize will have to increase by 60% over the next 40 years due to the growing world's population [1]. Additionally, a quarter of agricultural lands worldwide have suffered degradation, and there is a deepening awareness of the long-term consequences of a loss of biodiversity in terms of climate change. Oilseed crops, including maize, also play an important role in the agricultural economy. Globally, the demand for vegetable oils is increasing due to the increasing per capita consumption of oil in our daily diets and its use as biofuels [2]. By 2050, the global demand for vegetable oils is expected to be more than twice the current production. Though the need for maize crop is expected to increase, the crop productivity is limited by many abiotic and biotic stresses. A range of new technologies have been developed to enhance the productivity of this crop. Here, the current work presents an overview and discusses recent progresses on maize research that could open up new opportunities for crop improvement.

2. Technologies developed to enhance maize productivity

Molecular breeding approach in maize starts with identifying and validating quantitative trait loci (QTLs) linked to abiotic stress tolerance. Following the identification and validation of the markers associated with QTLs for traits of interest, the candidate QTLs or genes can be introgressed in elite lines through marker-assisted backcrossing. Over the past years, linkage mapping was used to identify QTLs [3]. However, association genetics is currently

used to enhance this work in numerous crops [4]. Nested association mapping is also being utilized for the genome-wide dissection of complex traits in maize crop [5]. Association mapping is highly recommended to be used for identifying traits associated with abiotic stresses [6]. Marker-assisted backcrossing has also been utilized for complex traits such as tolerance to drought, salinity, and heat, which are the key traits targeted for improving and developing crops that are adapted to low rainfall, salinity, and high temperature conditions. Marker-assisted backcrossing may not be an effective approach for introgressing QTLs in some cases. On the other hand, two other molecular breeding approaches, marker-assisted recurrent selection and genomic selection, can overcome this issue [7]. The genetic progress obtained using marker-assisted recurrent selection and genomic selection is greater than that can be obtained using marker-assisted backcrossing. Another technology for enhancing complex traits has been developed and is based on genome-wide selection. Although marker-assisted backcrossing and marker-assisted recurrent selection need provided QTL information for complex traits, information on marker trait associations is not necessarily needed for genome-wide selection [8]. Furthermore, genome-wide selection relies on the information associated with the prediction of the genomic-recorded breeding values of progeny.

Most areas planted with maize currently involve transgenic varieties, and the vast majority of hybrids are now resistant to insects and herbicides. Bt maize containing the protein cry-1fAb has been started to be grown in 2007 in order to control *Spodoptera frugiperda*. RR maize, which is resistant to glyphosate-based herbicides, was then used as an alternative for the management of weeds. Maize productivity relies on the genetic characteristics of the hybrid grown, environmental conditions, and the farming technology used [9]. The potential for the grain production may be affected by the interaction between the hybrid and the cultivation conditions. Cardoso et al. [10] recorded varying responses of cultivars being well-adapted to a wide range of conditions, in which they maintain their productivity. In conclusion, besides the potential use of biotechnological and genetic approaches in the improvement of different plant species [11–19], they could be used for improving maize yield and quality. The current work would shed light on the advancements made in those technologies.

Author details

Mohamed A. El-Esawi[1,2*]

*Address all correspondence to: mohamed.elesawi@science.tanta.edu.eg

1 The Sainsbury Laboratory, University of Cambridge, Cambridge, United Kingdom

2 Botany Department, Faculty of Science, Tanta University, Tanta, Egypt

References

[1] FAO. World Agricultural Production [Internet]. 2011. Available from: http://faostat.fao.org/default.aspx [Accessed: 2013-June-14]

[2] Samarth NB, Mahanwar PA. Modified vegetable oil based additives as a future polymeric material-review. Open Journal of Organic Polymer Materials. 2015;**5**:1-2

[3] Varshney RK, Tuberosa R. editors. Genomics-Assisted Crop Improvement: Genomics Approaches and Platforms. Vol. 1. The Netherlands: Springer; 2007

[4] Hall D. Using association mapping to dissect the genetic basis of complex traits in plants. Briefings in Functional Genomics. 2010;**9**:157-165

[5] Yu J et al. Genetic design and statistical power of nested association mapping in maize. Genetics. 2008;**178**:539-551

[6] Gupta PK et al. Linkage disequilibrium and association studies in plants: Present status and future prospects. Plant Molecular Biology 2005;**57**:461-485

[7] Tester M, Langridge P. Breeding technologies to increase crop production in a changing world. Science. 2010;**327**:818-822

[8] Jannink JL et al. Genomic selection in plant breeding: From theory to practice. Briefings in Functional Genomics 2010;**9**:166-177

[9] Tollenaa M, Lee EA. Yield potential, yield stability and stress tolerance in maize. Field Crops Research, Amsterdam. 2002;**75**:161-169

[10] Cardoso MJ, Carvalho HWL, Rocha LMP, Pacheco CAP, Guimaraes LJM, PEO G, Parentony SN, Oliveira IR. Identificação de cultivares de milho com base na análise de estabilidade fenotípica no Meio-Norte brasileiro. Revista Ciência Agronômica, Fortaleza. 2012;**43**:346-356

[11] El-Esawi MA. SSR analysis of genetic diversity and structure of the germplasm of faba bean (*Vicia faba* L.). Comptes Rendus Biologies. 2017;**340**:474-480

[12] El-Esawi M, Arthaut L, Jourdan N, d'Harlingue A, Martino C, Ahmad M. Blue-light induced biosynthesis of ROS contributes to the signaling mechanism of Arabidopsis cryptochrome. Scientific Reports. 2017;**7**:13875

[13] El-Esawi MA, Elkelish A, Elansary HO, et al. Genetic Transformation and Hairy Root Induction Enhance the Antioxidant Potential of *Lactuca serriola* L. Oxidative Medicine and Cellular Longevity, vol. 2017, Article ID 5604746, 8 pages. DOI:10.1155/2017/5604746

[14] El-Esawi MA, Elansary HO, El-Shanhorey NA, Abdel-Hamid AME, Ali HM, Elshikh MS. Salicylic acid-regulated antioxidant mechanisms and gene expression enhance rosemary performance under saline conditions. Frontiers in Physiology. 2017;**8**:716. DOI: 10.3389/fphys.2017.00716

[15] Vwioko E, Adinkwu O, El-Esawi MA. Comparative physiological, biochemical and genetic responses to prolonged waterlogging stress in okra and maize given exogenous ethylene priming. Frontiers in Physiology. 2017;**8**:632. DOI: 10.3389/fphys.2017.00632

[16] El-Esawi MA. Micropropagation technology and its applications for crop improvement. In: Anis M, Ahmad N, editors. Plant Tissue Culture: Propagation, Conservation and Crop Improvement. Singapore: Springer; 2016b. pp. 523-545

[17] El-Esawi MA. Nonzygotic embryogenesis for plant development. In: Anis M, Ahmad N, editors. Plant Tissue Culture: Propagation, Conservation and Crop Improvement. Singapore: Springer; 2016c. pp. 583-598

[18] El-Esawi MA. Somatic hybridization and microspore culture in *Brassica* improvement. In: Anis M, Ahmad N, editors. Plant Tissue Culture: Propagation, Conservation and Crop Improvement. Singapore: Springer; 2016d. pp. 599-609

[19] El-Esawi MA, Germaine K, Bourke P, Malone R. AFLP analysis of genetic diversity and phylogenetic relationships of *Brassica oleracea* in Ireland. Comptes Rendus Biologies. 2016b; **339**:163-170

Chapter 2

Polyembryony in Maize: A Complex, Elusive, and Potentially Agronomical Useful Trait

Mariela R. Michel, Marisol Cruz-Requena,
Marselino C. Avendaño-Sanchez,
Víctor M. González-Vazquez,
Adriana C. Flores-Gallegos, Cristóbal N. Aguilar,
José Espinoza-Velázquez and
Raúl Rodríguez-Herrera

Additional information is available at the end of the chapter

http://dx.doi.org/10.5772/intechopen.70549

Abstract

Polyembryony (PE) is a rare phenomenon in cultivated plant species. Since nineteenth century, several reports have been published on PE in maize. Reports of multiple seedlings developing at embryonic level in laboratory and studies under greenhouse and field conditions have demonstrated the presence of PE in cultivated maize (*Zea mays* L.). Nevertheless, there is a lack of knowledge about this phenomenon; diverse genetic mechanisms controlling PE in maize have been proposed: Mendelian inheritance of a single gene, interaction between two genes and multiple genes are some of the proposed mechanisms. On the other hand, the presence of two or more embryos per seed confers higher nutrimental quality because these grains have more crude fat and lysine than normal maize kernels. As mentioned above, there is a necessity for more studies about PE maize in order to establish the genetic mechanism responsible for this phenomenon; on the other hand, previous studies showed that PE has potential to generate specialized maize varieties with yield potential and grain quality.

Keywords: *Zea mays* L. polyembryony, genetic control, ploidy level, apomixis, xenia

1. Introduction

Polyembryony (PE) can be defined as the simultaneous emergence of two or more seedlings from one germinated seed [1]. The plant polyembryony phenomenon was discovered by Van

Leeuwenhoek in 1719 and reported in orange seeds and can be classified into two main types that are based on the cellular origin of embryogenesis either, gametophytic and sporophytic [2, 3].

This phenomenon occurs spontaneously in several plants species although at low frequencies. The term "polyembryony" also reports it as the division of one sexually produced embryo into many, and the resulting ones are genetically identical to each other, but distinct from their mother [4]. However, some PE versions have to feature of high potential with agronomical applications in maize [5]. This phenomenon is common in gymnosperms and less frequent in angiosperms [6]. Shukla in 2004 [7] reported about 59 families, 158 genera, and 239 vegetal species having this trait. Embryos in polyembryonic seeds may originate from embryo sac (ovule, zygote, synergids, and antipodes), nucellar tissue, or the integument [8, 9]. Therefore, may be monoploid (containing half (n) of the normal number of chromosomes), or diploid (with a normal number of chromosomes (2n)) [10–13].

Embryological studies in nineteenth and twentieth centuries demonstrated that the adventitious embryos present in a seed in addition to the sexual embryo can be formed based on different structures of ovule and embryo sac structures [14–18]. Maize PE has been studied for almost 100 years, judging from published reports [10, 19–29]. Although, this phenomenon has been studied by different authors, there are still many questions about the origin, causes, PE gene and its relationship with apomixis and pollen source, and the environmental effect on the expression of this feature [30]. This study provides a review of the present literature on this phenomenon, applications, and possible causes of PE and, particularly, discusses this phenomenon in maize.

2. PE in nature

PE has been reported in different plant species such as almond [31], citrus [32–34], mango [7, 35], peach [36], rice [37], soybean [38], strawberry [39], papaya [40], kiwi, apple [41], safflower [42], alfalfa [43, 44], lemon [45], grape [46], and olive cultivars [3]. Polyembryony was shown only in 8 of the 24 selected olive cultivars; this specificity of cultivar as in other fruit species agrees that polyembryony is also a genetically regulated character. The latter has two diploid (2n) embryos, one from zygote and the other from the nucellus [43]; potato and flax with two embryos, one diploid (2n) embryo from zygote and one haploid (n) from a synergid [47, 48]; wheat with two embryos in the same bag, an embryo of oosphere (n) and another from the fertilized (2n) synergid [49]; asparagus two diploid embryos from proembryo division [50]; citrus (*Citrus* spp.) with a normal embryo of sexual origin and others that develop from nucellus [31]; and papaya (*Carica papaya*) [40]. It has been assumed that the plants are of zygote origin, and there have been no genetic tests; occasionally, multiple embryos come from cultured ovules [46].

Most of the citrus cultivars are polyembryonic, for example, most lemon crops produce several embryos per seed, which is why it is necessary to rescue the zygotics, to reduce abortion and competition with nucleic embryos [45]. Polyembryony also has been reported among certain insects as parasitic wasps [51, 52] such as *Copidosoma floridanum* [53] and even mammals such as armadillo, which give birth to several offsprings, all twins [54]. Humans that originate in this way are the so-called identical twins, who are mostly genetically identical [55].

Polyembryonic, called embryo generation along with the zygotic embryo in a single seed, is widespread in angiosperms. The development of additional embryos may be induced by exogenous factors, such as pollen irradiation, higher temperatures, and herbicides, which are employed during and after flowering [56]. Polyembryony has been observed in sexual ferns and attributed to multiple fertilizations, and report this phenomenon in *Pteris tripartite* Sw. where they obtain from two to eight sporophytes, observed from a single gametophyte [57].

2.1. PE in maize

PE in maize is a phenomenon poorly studied. In addition, some research reports about this trait are contradictory. Sharman in 1942 [58] noted that a maize line had two embryos that emerged from a single caryopsis, whereby they were selected and dissected. The two embryos appeared to be completely separated except by the scutellum. This suggests that the twin characteristic showed up early and was probably caused by a longitudinal division or a constriction of the cell mass that was the stage of "pro-embryo." The above results suggested that both embryos were identical and produced typical plants with normal chromosome number 2n.

Morgan and Rappleye in 1951 [24] induced PE in maize after exposing pollen to different X-ray doses and crossing females of the same line with that pollen; after sowing the obtained seeds, it was observed that the presence of PE was up to 18% of the seeds. Thus, concluded that treatment with X-rays causes a significant deviation from the normal reproduction process resulting in the formation of numerous embryos. They also reported that double embryo seeds produced plants with different heights, indicating that haploid plants may occur among the polyembryonic lines, resulting from plant crosses where pollen was exposed to X-rays. Earlier reports mentioned maize with multiple plumules and primary roots, but with a single scutellum, concluding that these plants did not come from two embryos, but from one abnormal embryo [21]. This feature was also mentioned by Kempton in 1913 [19] and Weatherwax in 1921 [20] and was called false polyembryony. In all cases, there occurred two stems and two primary roots. Besides, a case was found where three stems were attached to a single cotyledon. After two generations of a line with this trait, it was observed that this peculiarity was lost.

Pešev in 1976 [25] reported the derivation of several inbred lines from a population that formerly exhibited a few twin plants; the inbred lines showed the twin condition in frequencies that ranged from 2.1 to 25.3%. Pollacsek in 1984 [59] reported that in the Old French INRA F1254 line, it was found that 4.5% of the plants were with double stems and determined that the nature of this trait was an early fasciation that takes place during embryogenesis. This trait with incomplete penetrance had low probably due to oligogenic control.

In 1973, the Instituto Mexicano del Maíz at Universidad Autónoma Agraria Antonio Narro (IMM-UAAAN) located in Saltillo, Mexico, generated a maize population which presented polyembryonic seeds with a frequency of 1.5%. This material was improved with a process of recurrent selection for 5 years under the assumption that this may lead to a gradual increase of favorable alleles for PE, and at the same time, maintain high genetic variability [5, 26]. To avoid that selection carried to inbreeding, twin crosses were made with elite inbred lines from a different origin [5]. This population in addition to genes for polyembryony had the brachytic two genes (*br2*). In 1991, 47% of the polyembryonic plants were observed in the population. Now, researchers decided to separate this population into two according to the phenotype in high or normal and brachytic (dwarf) plants. These have the brachytic2 (*br2*) gene [5]. The *br2* is a recessive gene that has an agronomic potential because it results in the shortening of the internodes of the lower stalk without an obvious reduction in other plant organs [60] that modulates polar auxin transport in the maize stalk. This gene encodes a protein similar to adenosine triphosphate (ATP)-binding cassette transporters of the multidrug resistant (MDR) class of P-glycoproteins (PGs) [61].

Four years later, the percentage of PE in both populations averaged 60%; the most common issue was found in seeds with double seedlings, but the number of seedlings per seed was as high as six (**Figure 1**). In 1996, each population (normal and brachytic plant height) was divided into two subpopulations, one with plants where PE frequency was high and one with plants where PE was low, having four different populations: the normal height plant and high polyembryony (NAP); normal height plant and low polyembryony (NBP); brachytic plant height and high polyembryony (BAP); and dwarf plant height with low polyembryony (BBP). In 1998, dwarf and normal populations reached 61 and 63% of PE, respectively [5]. The frequency of PE is currently 65 and 60% for the dwarf and normal populations, respectively;

Figure 1. Polyembryonic and nonpolyembryonic maize seedlings. (a) Left to right: Normal maize phenotype, twin maize (PE maize) seedling both normal and twin of 21 days old, triple, and quadruple maize seedling of 28 days old. (b) Sixfold seedling: multiple seedling almost independent, at least sharing scutellum. (c) Several ways in which twins seedlings are observed; there are also cases of two or more radicles per PE plants. *Photographs provided by Jose Espinoza-Velazquez IMM-UAAAN.*

the higher frequency of PE in these populations are twin plants (**Figure 1**), followed by triple and presenting uncommon seedlings–quadruple, quintuple, and sextuple [28]. Espinoza-Velazquez and Vega in 2000 [62] worked with subpopulations of IMM-UAAAN and reported that in the period 1995–2000 the selection for the PE has gained between 2 and 3% per cycle. They led the polyembryonic populations to levels above 60% PE, while the reverse selection (contrary to PE) groups rapidly leads to frequencies less than 6% PE.

3. Agronomic benefits from PE

Polyembryonic seed is an important feature due to commercial multiplication [63]. Citrus has a normal embryo of sexual origin and others that develop from ovule nucellus, so all these embryos from nucellus are identical to the parent plant so that they may be used as rootstocks by their rusticity and uniformity [64]. The PE is an extremely rare phenomenon in maize; however, this trait may confer great benefits since in this case, plants may have increased production and competitiveness because a seed may produce two to six normal plants favoring production because of the increase of number of plants and ears per surface unit [30].

Other benefits are lower production costs because with the same number of seeds, farmers can have more plants per unit area. So to plant a unit area will require less seed that will result in lower storage and transportation costs [5]. However, yield performance and population density experiments are needed to evaluate the improvement in grain yield because of polyembryonic maize varieties.

4. Nutritional benefits from PE

Pešev in 1976 [25] reported a significant increase in protein (4.5%), lysine g/100 g dry material (38–70.9%), lysine g/100 g protein (21.3–34.0%), and oil (3.5–13.6%) in polyembryonic maize grains compared to those with a single embryo. Other authors have reported a positive increase in polyembryonic maize dough, detecting a positive association between PE and oil content (22% higher than a native variety) with a high percentage of unsaturated oils and a better relationship between oleic and linoleic acids. The average of crude protein in polyembryonic maize is 10% and was 8% higher than a native variety. The crude fat content (FC) of grain in NAP and BAP populations showed an overall average of 6.2% [65]. This may be attributable to the positive correlation between PE and lipid concentration in the grain. FC quantitative superiority of maize PE may also be more qualitative because from 55 to 65% of the grains of an ear has two or more embryos [28]. This suggests that selection in favor of polyembryony increases indirectly grain content of nutriments as crude fat and lysine; a condition that could be exploited in the design of new varieties of PE maize, combining high yield and grain quality.

Gonzalez and collaborators [28] in a study on nutritional quality and quantity of PE grains derived from crosses between the IMM-UAAAN-BAP population (PE) and Tuxpeño Population high oil content (HOC) of Centro International de Mejoramiento de Maíz y Trigo (CIMMYT), to generate PE:HOC germplasm, direct and reciprocal crosses, as well as backcrossing to both parents, were performed. The authors obtained the following germplasm combinations (0:100, 12.5:87.5, 25:75, 37.5:62.5, 50:50, 62.5:37.5, 75:25, 87.5:12.5, and 100:0). These authors noted that crude fat content (CF) and lysine (Lys) may be raised increasing the doses of HOC and PE, respectively. The optimal combinations of germplasm PE:HOC for nutritional grain quality combinations were 50:50 (Lys = 2.7%; FC = 6.9%); these values were higher than those observed in common maize. The PE present in BAP population induced the highest value for lysine (4%). The PE in maize may be usable as an alternate route in the designing of varieties for special applications. In addition to the pattern for potential yield, the nutritional value of the grain, increasing quantity, and quality of protein and oil, which under the hypothesis that two or more embryos per seed, will increase the storage capacity of quality nutriments [66]. Cruz [67] studied the chemical, physical, and rheological properties of dough, tortilla, and grain of maize populations with high polyembryony. They concluded that the physical and chemical characteristics of polyembryonic maize are within those acceptable ranges for the production of food products, such as tortillas and flour.

5. Types of polyembryony

Analysis of different classifications of PE has shown that the main criterion for classification includes the origin of the initial cell, embryo formation pathways, and their genetic characteristics. The first classification system was proposed by Braun [68], who described four possible routes for the formation of adventitious embryos as a result of a merger of two or more eggs, developing several embryo sacs in the same ovule, or as result of a pro-embryonic division. According to Lakshmanan and Ambegaokar in 1984 [69], the PE is classified into "simple" or "multiple," depending on the presence of one (single) or more (multiple) embryo sacs in the same ovule and events that can occur in both types. In angiosperms, after the first mitosis, the zygote is divided into two and then forms an embryo of each of the parts. It may also happen that the nucellus is divided into several parts from which originate many embryo sacs. Sometimes, only one of them is fully developed. In such a case, the seed embryo is formed from the union of gametes. This is a reproduction mode called apomixis that is a common event among flowering plants and is identified only by careful genetic study because the seeds look normal [70]. The PE may arise in angiosperms in four different ways: (1) PE for "cleavage" or division of the embryo to form more than one, (2) by the formation of embryos from different embryo sac cells to the egg cell, (3) by the development of more than one embryo sac within the same ovule derivative thereof from the megaspore mother cell or cells of the nucellus, and (4) by activation of a somatic cell or sporophytic ovule to form the embryo [71].

The PE per cleavage generates embryos from the zygote and sometimes from its suspensor within the embryo sac [69]. The synergids are the most common cells within the embryo sac that can form embryos, which can be fertilized by the egg cell and by the same pollen tube. However, in the absence of fertilization of the polar nuclei, the endosperm is not formed and the entire process collapses. Moreover, the process may also involve several pollen tubes to fertilize the egg cell, polar nuclei, and synergids, achieving normal endosperm development [71]. Antipodal embryos are rare. Some authors question the possibility of forming embryos from antipodal cells [72]. Although, it has been observed that the number of seedlings per seed under greenhouse conditions of polyembryonic maize can be up to six (**Figure 1**), which approaches to the number of nuclei in the embryo sac. More studies are needed to elucidate the origin of PE in maize.

Embryos formed from the sporophytic cell (2n) are known as adventitious, and they are generated from the nucellus and integuments. In nucellar polyembryony, cells generally contain a starchy and dense cytoplasm, they actively divide and become embryonic masses directing their way into the embryo sac, and cell activation may be stimulated in an autonomous way or by pollen tube inserted into the sac or even by pollination. The angiosperms that are distinguished by nucellar polyembryony are *Citrus* and *Mangifera* [71]. Batygina and Vinogradova in 2007 [2] classified the PE into two main types: gametophytic and sporophytic. The first type is a PE related with the phenomenon associated with the formation of the adventitious embryo gametophytic cell: synergids and antipodal, also as an embryonic cell when the embryonic sac is developed further. While PE sporophyte is characterized by the development of adventitious embryos from sporophytic cells: mother (integumental and nucellar polyembryony) or daughter (polyembryony monozygotic twins).

In maize, it has been suggested that the PE is of suspensory type [8] as well as the zygotic type as described by Lakshmanan and Ambegaokar [69], where the embryos are arising spontaneously in suspensor cells from zygotic embryo. Erdelska [9] in a histological analysis (1996) suggests that PE is produced according to the origin of the embryos, their location in the grain (caryopsis), difference in structure (common tissues), and type of germination. From these concepts, the PE can originate in three ways: (1) two embryo sacs multi-embryonic commonly are located on opposite sides, or distance in the grain, which lack of common tissues and germinate independently, (2) cases of twins or triplets coming from individual egg cell, or embryo sac cells with multi-egg capabilities that are closely adhered, but separated by epidermal layers, with an endosperm in common and independent radicles and plumules, and (3) polyembryos arising from multiplication of egg cell cleavage spontaneously or after any induction, which share a common suspensors that are part of scutellum and radicle surface layers and due to this reason, embryos germinate with separated plumules but one root complex.

Moreover, in a study on morphology and anatomy of maize radicles as well as frequency of seedlings and multiple radicles per germinated seed, performed using two maize populations from the IMM-UAAAN polyembryonic germplasm as well as their direct and reciprocal crosses with Non-PE genotypes, it was found that the PE and multiple radicles trait occurred only in the progeny of the two polyembryonic populations and the hybrids between them. Some PE seedlings presented simultaneously multiple radicles, whereas other PE seedlings do not show these multiple radicles that were observed in variable number and conformation. In some cases from two to four roots, separated or merged with some degree of histological level, including the vascular cylinder. The average frequency of PE and multiple radicles was 62 and 14%, respectively [29]. This can be explained as a phenomenon of cleavage polyembryony by affecting cell division, making proembryonic form various embryonic axes that are attached by certain structures [73].

6. Possible causes of polyembryony

Despite the interest in the factor that may induce and affect the frequency of PE in different species, PE mechanisms and causes are not yet entirely clear. It is considered that the causes of PE are mainly genetic, although there is a strong environmental component in PE expression. One of the first suggested causes of PE was a hormonal imbalance [74], although recently PE has been attributed to genetic causes such as meiotic and/or mitotic chromosome irregularities and polyploidy hybridization. Polyembryony (polymeric embryos) can develop spontaneously in different plants with live flowers or can be induced in situ by various treatments, such as synthetic auxins, X-rays, or inhibitors of auxin polar transport [75]. Of these three, the most reported are irregularities during the meiotic and/or mitotic process, which are governed by the *ig* gene. It has been reported particularly in maize that the presence of multiple embryonic cells due to the mutant *ig* gene affects the number of mitotic divisions. However, there are not sufficient studies to ensure that mutation of this gene is associated with PE. On the other hand, it has been reported that PE can be increased by a selection or delayed pollination [8] suggesting a genetic component.

There are controversial reports on the genetic nature of PE. Shukla in 2004 [7] studied the genetic diversity of polyembryonic and monoembryonic mango and found that the two phenomena have a different genetic basis. Similar results were obtained by Andrade-Rodriguez [76] who used RAPD markers for identification of zygotic and nucellar seedlings in polyembryonic *Citrus reshni* and reported that it was possible to identify both types of seedlings. By contrast, Martínez-Gómez and Gradziel [31] analyzed the genetic structure of almond seedlings from mono- and polyembryonic seeds and found that the seedlings have a similar genetic composition in both types of embryos. It was also mentioned that variation of polyembryony may be affected by a type of pollinator, available pollen amount, plant nutrition, environment temperature, soil moisture and temperature, and air velocity. Therefore, factors affecting pollination or fertilization of seed development will also affect PE percentage and number of embryos per seed [77].

The occurrence of PE varies greatly and is influenced by environmental conditions. Plants from the same polyembryonic seed often are viable, although some of the plants may show a weak development of their leaves [31]. Andrade Rodríguez in 2005 [76] found that the environmental conditions during the growing season of *Citrus volkameriana* affected PE frequency; in addition, the fruit morphological characteristics do not indicate the PE frequency. These authors determined that the zygotic lines have a different RAPD pattern to nucellar lines and found that only 25.9% of the polyembryonic and 87.5% of the monoembryonic plants are of sexual origin and that in the polyembryonic seeds not all zygotic embryos were produced by the small embryos located in the micropyle.

There are reports where PE in maize was induced by treating the developing caryopses with 2,4-dichlorophenoxyacetic acid (2,4-D), on the second day after pollination finding that about 40% of the seeds were polyembryonic. The same authors also observed that polyembryonic caryopses were smaller than normal because of lower growth potential [8].

http://dx.doi.org/10.5772/intechopen.70549

6.1. Polyembryony and apomixes

Webber [1] noted that many cases of adventitious cell formation in angiosperms are related with apomixes, and it is very likely that PE and apomixis can be interconnected. Apomixis in *Citrus* is known as polyembryony because multiple somatic embryos are developed simultaneously with the zygote embryo in the seed [78].

Genes that initiate and control apomixis will lead to the development of true reproductive hybrids for the genotype of a superior hybrids; apomixis can be divided into different categories: (i) adventitia or sporophytic type is where the embryos differ from the somatic cells in the eggs without the formation of megagametophyte; (ii) apospory, where the megagametophyte is to be developed from a somatic cell within the ovum; and (iii) diplospory with the development of megagametophytes of a nonreduced miaspore stem cell. Apomictic processes mimic many of the events of sexual reproduction to give rise to seeds without fertilization [79]. However, polyembryony has been characterized as the occurrence of more than one embryo in a seed, polyembryony in angiosperms may appear by excision of the proembryo, or formation of embryos by the cells of the embryonic sac [37].

Some varieties of citrus express a form of apomixis nucellar embryo in which adventive, the embryos are developed from the nucellus embryonic sac tissue. This feature appears in many seeds containing multiple embryos (polyembryony) [80]. Different species present several reproductive traits that appear to be interacting in the generation of PE. Gupta in 1996 [81] reported in guggul (*Commiphora wightii*) the occurrence of apomixis not pseudogamous (development of an embryo only from maternal chromosomes after activation of the egg by a sperm: sperm penetrates the egg, causes division, but there is no effective fertilization), nucellar PE and autonomous endosperm formation suggesting that plants have reproductive and survival strategies in the absence of male plants, but in the presence of males, sexual reproduction can occur. Moreover, in 2005, Mendes-Rodriguez [82] studied *Eriotheca pubescens*, which presents apomixis and adventitia polyembryony, found that in seeds, the zygote became a sexual embryo simultaneously with apomictic adventitious embryos that developed from nucellus cells. The adventitious embryo developed more rapidly than sexual ones, but they are morphologically similar so that 44 days after anthesis it was impossible to distinguish the sexual from the apomictic embryos.

Espinoza-Velazquez and De Leon in 2005 [83] asserted that maize populations might contain the ability to manifest asexual reproduction by seed, some form of apomixis. They were based on the history of polyploidy and polyembryony in the IMM-UAAAN populations and preliminary work on atypical reproductive behavior in maize. The introduction of apomixis in maize has been attempted through conventional backcrossing, using *Tripsacum* species as the source, from where can be generated viable seeds from intergeneric hybridization, which were produced in an apomictic way when they were pollinated using common maize [84]. This suggests that pollen source can influence apomictic embryo development. However, despite the effort to introgress apomixis into maize from its wild relative *Tripsacum dactyloides*, the attempts to generate apomictic maize have failed so far. As Leblanc [85] have concluded that "epigenetic information imposes constraints for apomictic seed development and seems pivotal for a transgenerational propagation of apomixis."

Several studies have discussed the evolution of apomixis and adventitious embryos on the subject of their similarity in regard to asexual propagation [86–89]. Given that there is a lack of clear distinction between PE, apomixis, and adventitious embryos, this is assumed because of all these phenomena have similarities in asexual reproduction. However, PE is distinguished from the other two processes on the basis of its requirement of sexual reproduction and genetic composition of their offspring. Since there is a clear distinction between PE, apomixis, and adventitious embryos, all of the above is assumed to have similarities in asexual reproduction.

7. Polyembryony and pollen source

In maize, various experiments have been conducted to show an effect of the origin and nature of pollen on grain development. This has been expressed as the difference in weight between the grains of selfing and those of cross-fertilization, where the grain weight of cross-fertilization increased 10.1% [21]. From the genetic point of view, the advantage of cross-fertilization can be interpreted in terms of complementarity among genes from male and female by some enzymatic systems in terms of heterosis [90]. The effect of pollen source has been reported affecting seed composition. In the case of QPM (high-quality protein maize), if normal maize pollen fertilizes QPM female plants, essential amino acid content in the grains is decreased; in the case of lysine, it is up to 30% by which the maize grains from QPM plants reach a protein quality similar to normal maize [91].

Villarreal in 2010 [30] conducted a study using 16 samples of maize grains, a product of crosses among four female and four different male lines (**Table 1**). He found a higher percentage of PE in the offspring of females with high PE levels crossed with a polyembryonic and genetically

	Line description	Coding
Female lines	1. Normal height and high polyembryony	NAP
	2. Brachytic line with high polyembryony	BAP
	3. Normal height and low polyembryony	NBP
	4. Brachytic line with low polyembryony	BBP
Male lines	1. Polyembryonic line and genetically related to female lines	PERE
	2. Polyembryonic line and genetically unrelated to female lines	PENORE
	3. Nonpolyembryonic line that is genetically related to female lines	NOPERE
	4. Nonpolyembryonic line that is genetically unrelated to female lines	NOPENORE

Table 1. Female and male polyembryonic and nonpolyembryonic maize lines.

unrelated male, compared to when the same female was crossed with a polyembryonic and genetically related male. These results suggest a possible genetic complementation conditioning maize PE and some possible maternal effects as well.

8. Genetic studies

Regarding genetic control, PE in maize has been reported as a trait of simple recessive inheritance [10, 13, 92], as well as a quantitative inheritance [5, 25, 26]. According to these authors, the manifestation of this character can arise from major effect genes (monogenic nature) or polygenes (quantitative nature). In the first type, one needs to emphasize the role of *ig* gene, which in a homozygote recessive condition generates in seeds with a monoploid embryo in 3% of the cases and in 6% PE [11], or by an unidentified recessive gene, as noted by Pilu [92]. However, Pešev [25], Rodriguez, and Castro [93] and Castro [26], cited by Espinoza [5], mentioned that inheritance of PE is quantitative, and the latter authors note that PE which they worked presented a heritability of 65%, calculated by the method of midparent-offspring regression method. There is evidence that maize PE has a heritable basis of a quantitative nature; however, inconsistent behavior, regarding fixing PE in genetic groups, suggests involvement of other genetic and reproductive phenomena such as nucleus-cytoplasm interaction and reductional type parthenogenesis. Microarrays and SSH have been used to identify the genes associated with polyembryony in *Citrus*. Studies have also been made to associate polyembryonic with heat stress [94].

As reported by Puri, polyembryony in rice is caused by insertion of mutagenesis, where they employ molecular tools for the cloning of the polyembryo gene (Ospe) in Basmati 370, and mention that for the F3 population, the polyembryony was not segregated with the expected proportion, suggesting that there is variable penetrance and expressiveness for the mutant. Penetration is related when a phenotype is expressed for a particular genotype, which expressively refers to the degree to which a phenotype is expressed after penetrance, obtaining polyembryonic seeds of twins, triplets, and rare quadruplets that varied from 9.8 to 21.8% [95].

A study about the combination of PE germplasm with a nonpolyembryonic (Non-PE) source indicates a masking of PE trait in the F_1 generation in the crosses of the polyembryonic populations (NAP and BAP) with the Tuxpeño population that has high oil content and belongs to the CIMMYT collection [28]. Continuing with this experimental line, Espinoza Velazquez [96] reported on the probable genetic mechanisms involved in the PE expression. After analyzing the observed PE frequencies in the F_2 and RC_1 generation, they found that PE frequency did not agree on the expected in the case of a recessive gene but to the two interacting loci with epistasis of the kind of 15:1 double recessive for PE. A more recent study on PE reported by Musito Ramirez in 2008 [97] who worked with S_1 inbred lines derived from the NAP population (**Table 1**) found that inbreeding of S_1 lines did not increase PE frequency. Moreover, Espinoza-Velazquez in 2012 [29], after performing a histological study of 3-day-old radicles, belonging to genotypes derived from crossings among the NAP and BAP populations (**Table 1**) with the Tuxpeño HOC population, found that PE frequency and multiple radicles (two or more roots per seed) were 60 and 14%, respectively; however, the traits were masked in the F_1 hybrids, manifesting that the PE as a recessive trait. Rebolloza in 2011 [27], who worked with BAP and NAP maize populations (**Table 1**), found that PE showed a Mendelian inheritance pattern by the action of two loci, with epistatic interaction of duplicate recessive type having a F_2 segregation of 15:1, with an incomplete penetrance of a range from 20 to 50%; thus, according to this source, the exotic germplasm with PE is being crossed. These findings corroborate the proposed inheritance mechanism suggested by Espinoza-Velazquez [96].

9. Future trends

Maize PE is a trait that has different practical applications. As demonstrated by several authors, polyembryonic maize contains higher grain nutritional quality which allows to develop PE varieties with high fat content (6.5%) and lysine (4%) [28, 30] also crosses between PE and non-PE genotypes produces hybrids that do not express the PE trait because of its recessive genetic condition, but fat and lysine in the grain remain high, which may help to generate hybrids with higher grain nutritional quality [28]. In the case of studies that attempt to explain the PE in maize, it is necessary to apply the advances in molecular biology for the identification of the genes that are involved in the control of this trait, and if it is possible to sequence these genes in order to provide greater information of this trait and increase its agricultural utility by inserting such genes in lines with high agronomic potential or for further molecular studies on PE and its relation to polyploidy, xenia, and apomixis.

10. Conclusion

Polyembryony in maize has been documented first by the presence of multiple plants simultaneously born from a seed and cytological studies that have confirmed this trait. The type of PE inheritance could be governed by major genes or genes of a quantitative nature. Besides, the presence of two or more embryos per seed gives an advantage to these genotypes for higher grain nutrimental quality. However, more studies are required in order to fully understand the PE nature and control. On the other hand, reported studies showed that PE could be a useful trait in developing specialized varieties with yield potential and grain quality. However, there is not much molecular evidence that can help to fully understand the polyembryony trait.

Acknowledgements

This project was financially supported by Universidad Autónoma de Coahuila and Universidad Autonoma Agraria Antonio Narro. Thank the Mexican Council of Science and Technology (CONACYT) for the financial support during their graduate studies.

Appendices

CIMMYT	Centro International de Mejoramiento de Maíz y Trigo
FC	Fat content
HOC	High oil content
IMM-UAAAN	Instituto Mexicano del Maíz of the Universidad Autónoma Agraria Antonio Narro
Non-PE	Non-polyembryonic
PE	Polyembryony, polyembryonic
QPM	High quality protein maize
RAPD	Random amplified polymorphic DNA

Author details

Mariela R. Michel[1], Marisol Cruz-Requena[1], Marselino C. Avendaño-Sanchez[2], Víctor M. González-Vazquez[1], Adriana C. Flores-Gallegos, Cristóbal N. Aguilar[1], José Espinoza-Velázquez[2] and Raúl Rodríguez-Herrera[1]*

*Address all correspondence to: raul.rodriguez@uadec.edu.mx

1 Food Research Department, School of Chemistry, Universidad Autónoma de Coahuila, Saltillo, Coahuila, Mexico

2 Mexican Maize Institute, Universidad Autónoma Agraria Antonio Narro, Buenavista, Coahuila, Mexico

References

[1] Webber JM. Polyembryony. The Botanical Review. 1940;**6**:575-598

[2] Batygina TB, Vinogradova GY. Phenomenon of polyembryony. Genetic heterogeneity of seeds. Russian Journal of Developmental Biology. 2007;**38**:126-151. DOI: 10.1134/S1062360407030022

[3] Trapero C, Barranco D, Martín A, Díez CM. Occurrence and variability of sexual polyembryony in olive cultivars. Scientia Horticulturae. 2014;**177**:43-46. DOI: 10.1016/j.scienta.2014.07.015

[4] Craig SF, Slobodkin LB, Wray GA, Biermann CH. The 'paradox' of polyembryony: A review of the cases and a hypothesis for its evolution. Evolutionary Ecology. 1997;**11**:127-143

[5] Espinoza Velázquez J, Vega MC, Navarro E, Burciaga GA. Poliembrionía en maíces de porte normal y enano. Agronomía Mesoamericana. 1998;**9**:83-88

[6] Maheswari P, Sachar RC. Polyembryony. In: Maheswari P, editor. Recent Advances in the Embryology of Angiosperms. New Delhi: International Society of Plant Morphologists; 1963. p. 265-296

[7] Shukla M, Babu R, Mathur VK, Srivastava DK. Diverse genetic bases of Indian polyembryonic and monoembrionic mango (*Mangifera indica* L) cultivars. Current Science. 2004;**87**:870-971

[8] Erdelská O, Vidovencová Z. Cleavage polyembryony in maize. Sexual Plant Reproduction. 1992;**5**:224-226. DOI: 10.1007/BF00189815

[9] Erdelská O. Polyembryony in maize: Histological analysis. Acta Societatis Botanicorum Poloniae. 1996;**65**:123-125

[10] Kermicle JL. Androgenesis conditioned by a mutation in maize. Science. 1969;**166**:1422-1424. DOI: 10.1126/science.166.3911.1422

[11] Hallauer AR, Miranda FO. Quantitative Genetics in Maize Breeding. USA: Iowa State University Press; 1988. p. 468. DOI: 10.1007/978-1-4419-0766-0_6

[12] Huang BQ, Sheridan WF. Embryo sac development in the maize indeterminate gametophyte1 mutant: Abnormal nuclear behavior and defective microtubule organization. The Plant Cell. 1996;**8**:1391-1407. DOI: 10.1105/tpc.8.8.1391

[13] Evans MMS. The indeterminate gametophyte 1 gene of maize encodes a LOB domain protein required for embryo sac and leaf development. The Plant Cell. 2007;**19**:46-62. DOI: 10.1105/tpc.106.047506

[14] Litz RE, Schaffer B. Polyamines in adventitious and somatic embryogenesis in Mango (*Mangifera indica* L.). Journal of Plant Physiology. 1987;**128**:251-258

[15] Teppner H. Adventitious embryony in *Nigritella* (Orchidaceae). Folia Geobotanica & Phytotaxonomica. 1996;**31**:323-331

[16] Xi-mei D, Xu Y, Qun-ce H, Guang-yong Q. Observation on double fertilization and early embryonic development in autotetraploid polyembryonic rice. Rice Science. 2009;**16**:124-130. DOI: 10.1016/S1672-6308(08)60068-2

[17] Bittencourt NS Jr, Giorgi Moraes CI. Self-fertility and polyembryony in South American yellow trumpet trees (*Handroanthus chrysotrichus* and *H. ochraceus*, Bignoniaceae): A histological study of postpollination events. Plant Systematics and Evolution. 2010;**288**:59-76. DOI: 10.1007/s00606-010-0313-2

[18] Firetti-Leggieri F, Lohmann LG, Alcantara S, Ribeiro da Costa I, Semir J. Polyploidy and polyembryony in *Anemopaegma* (Bignonieae, Bignoniaceae). Plant Reproduction. 2013;**26**:43-53. DOI: 10.1007/s00497-012-0206-3

[19] Kempton JH. Floral Abnormalities in Maize. Bulletin U.S. Bureau of Plant Industry; 1913 No. 278

[20] Weatherwax P. Anomalies in maize and its relatives-I. Bulletin of the Torrey Botanical Club. 1921;**48**:253-255

[21] Kiesselbach TA. False polyembryony in maize. American Journal of Botany. 1926;**13**:33-36. DOI: 10.2307/2435405

[22] Randolph LF. Developmental morphology of the caryopsis in maize. Journal of Agricultural Research. 1936;**53**:881-916

[23] Skovsted A. Cytological studies in twin plants. Comptes Rendus Des Travaux Du Laboratoire Carlsberg. Serie Physiologique. 1939;**22**:427-446

[24] Morgan DT, Rapleyye RD. Polyembryony in maize and lily, following X-radiation of the pollen. Journal of Heredity. 1951;**42**:91-93. DOI: 10.1093/oxfordjournals.jhered.a106172

[25] Pešev NR, Petrović L, Zečević J, Milošević M. Study of possibility in raising maize inbred lines with two embryos. Theoretical and Applied Genetics. 1976;**47**:197-201. DOI: 10.1007/BF00278378

[26] Castro Gil M. Estudios sobre herencia y valor nutritivo de semillas con doble embrión. In: Avances de investigación. Sección Maíz. Universidad Autónoma Agraria Antonio Narro; 1979. p. 24-25

[27] Rebolloza-Hernández H, Espinoza-Velázquez J, Sámano-Garduño D, Zamora Villa VM. Polyembryony inheritance in two expperimental maize populations. Revista Fitotecnia Mexicana. 2011;**34**:27-33

[28] González-Vázquez VM, Espinoza-Velázquez J, Mendoza-Villarreal R, De León-Castillo H, Torres Tapia MA. Characterisation of maize germoplasm that combines a high oil content and polyembryony. Universidad y Ciencia. 2011;**27**:1-11

[29] Espinoza-Velázquez J, Valdés-Reyna J, Alcalá-Rodríguez JM. Morfología y anatomía de radículas múltiples en plántulas de maíz derivadas de cariopsis con poliembrionía. Polibotánica. 2012;**33**:207-221

[30] Villarreal A, Rodríguez Herrera R, Reyes Valdés MH, Espinosa Velázquez J, Castillo Reyes F. Xenia y su relación con la poliembrionía en maíz. Acta Química Mexicana. 2010;**2**:1-7

[31] Martínez-Gómez P, Gradziel TM. Sexual polyembryony in almond. Sexual Plant Reproduction. 2003;**16**:135-139. DOI: 10.1007%2Fs00497-003-0180-x

[32] Koltunow AM, Hidaka T, Robinson SP. Polyembryony in citrus. Accumulation of seed storage proteins in seeds and in embryos cultured *in vitro*. Plant Physiology. 1996;**110**:599-609. DOI: 10.1104/pp.110.2.599

[33] Aleza P, Juárez J, Ollitrault P, Navarro L. Polyembryony in non-apomictic citrus genotypes. Annals of Botany. 2010;**106**:533-545. DOI: 10.1093/aob/mcq148

[34] Nakano M, Shimada T, Endo T, Fujii H, Nesumi H, Kita M, Ebina M, Shimizu T, Omura M. Characterization of genomic sequence showing strong association with polyembryony

among diverse *Citrus* species and cultivars, and its synteny with *Vitis* and *Populus*. Plant Science. 2012;**183**:131-142. DOI: 10.1016/j.plantsci.2011.08.002

[35] Aron Y, Czosnek H, Gazit S. Polyembryony in Mango (*Mangifera indica* L.) is controlled by a single dominant gene. HortScience. 1998;**33**:1241-1242

[36] Zdruikovskaya-Rikhter, A. I. (1970). The phenomenon of polyembryony in the peach. Trudy Gosudarstvennogo Nikitskogo opytnogo Botanicheskogo Sada.1970:**46**:173-177

[37] Mu X, Jin B, Teng N. Studies on the early development of zygotic and synergid embryo and endosperm in polyembryonic rice ApIII. Flora-Morphology, Distribution, Functional Ecology of Plants. 2010;**205**:404-410. DOI: 10.1016/j.flora.2009.12.023

[38] Kenworthy WJ, Brim CA, Wernsman EA. Polyembryony in soybeans. Crop Science. 1973;**13**:637-639. DOI: 10.2135/cropsci1973.0011183X001300060015x

[39] Lebegue A. Recherches sur le polyembryonie les Rosacees (g. *Fragaria* et Achemilla). Bulletin de la Société Botanique de France. 1952;**96**:38-39

[40] Vegas A, Trujillo G, Sandrea Y, Mata J. Apomixis, poliembrionía somática y cigotica *in vivo* en lechosa. Interciencia. 2003;**28**:715-718

[41] Liznev V. N., Burdasov V. M. Polyembryony in apple trees. Bot. Z. 1971;**56**:416-421

[42] Vrijendra S, Deshpande MB, Nimbkar N. Polyembryony in safflower and its role in crop improvement. In: Proceedings of the VIth International Safflower Conference, İstanbul-Turkey, 6-10 June, 2005. SAFFLOWER: A Unique Crop for Oil Spices and Health Consequently, a Better Life for You; 2005. p. 1-20

[43] Bishop DS. *Polyembryony* and breeding behavior of polyhaploid twins in *alfalfa* MSc. [thesis]. Kansas State University; 1958

[44] Uzelac B, Ninkovic S, Smigocki A, Budimir S. Origin and development of secondary somatic embryos in transformed embryogenic cultures of *Medicago sativa*. Biologia Plantarum. 2007;**51**:1-6. DOI: 10.1007/s10535-007-0001-4

[45] Pérez-Tornero O, Porras I. Assessment of polyembryony in lemon: Rescue and in vitro culture of immature embryos. Plant Cell, Tissue and Organ Culture. 2008;**93**:173-180. DOI: 10.1007/s11240-008-9358-0

[46] Durham RE, Moore GA, Gray DJ, Mortensen JA. The use of leaf GPI and IDH isozymes to examine the origin of polyembryony in cultured ovules of seedless grape. Plant Cell Reports. 1989;**7**:669-672. DOI: 10.1007/BF00272057

[47] Frandsen NO. Polyembryony in potato. Zeitschrift fur Pflanzenzuchtung. 1968;**59**:343-358

[48] Green AG, Salisbury PA. Inheritance of polyembryony in flax (*Linum usitatissimum*). Canadian Journal of Genetics and Cytology. 1983;**25**:117-121. DOI: 10.1139/g83-023

[49] Sharmeen F, Rahman L, Alam MS, Shamsuddin AKM. Variation in the induction of haploidy and polyembryony in spring wheat through wheat × maize system. Plant Tissue Culture. 2005;**15**:7-14

[50] Rick CM. A cytogenetic study of polyembryony in *Asparagus officinalis* L. American Journal of Botany. 1945;**32**:560-569

[51] Grbic M. Polyembryony in parasitic wasps: Evolution of a novel mode of development. International Journal of Developmental Biology. 2003;**47**:633-642

[52] Segoli M, Bouskila A, Harari AR, Keasar T. Developmental patterns in the polyembryonic parasitoid wasp *Copidosoma koehleri*. Arthropod Structure & Development. 2009;**38**:84-90. DOI: 10.1016/j.asd.2008.05.003

[53] Zhurov V, Terzin T, Grbić M. (In)discrete charm of the polyembryony: Evolution of embryo cloning. Cellular and Molecular Life Sciences. 2007;**64**:2790-2798

[54] Loughry WJ, Prodöhl PA, McDonough C, Avise J. Polyembryony in armadillos. American Scientist. 1998;**86**:274-279

[55] Edwards RG, Mettler L, Walters DE. Identical twins and in vitro fertilization. Journal of In Vitro Fertilization and Embryo Transfer. 1986;**3**:114-117. DOI: 10.1007/BF01139357

[56] Pershina LA, Rakovtseva TS, Belova LI, Devyatkina EP, Silkova OG, Kravtsova LA, Shchapova AI. Effect of rye *Secale cereale* L. chromosomes 1R and 3R on polyembryony expression in hybrid combinations between (*Hordeum* L.)-*Triticum aestivum* L. alloplasmic recombinant lines and wheat-rye substitution lines T. *aestivum* L.-S. *cereale* L. Russian Journal of Genetics. 2007;**43**:791-797. DOI: 10.1134/S1022795407070113

[57] Ravi BX. In vitro polyembryony induction in a critically endangered fern, *Pteris tripartita* Sw. Asian Pacific Journal of Reproduction. 2016;**5**:345-350. DOI: 10.1016/j.apjr.2016.06.012

[58] Sharman BC. A twin seedling in *Zea mays* L. twinning in the gramineae. New Phytologist. 1942;**41**:125-129. DOI: 10.1111/j.1469-8137.1942

[59] Pollacsek M. Twin stalks in maize. Maize Genetics Cooperation Newsletter. 1984;**58**:56

[60] Pilu R, Cassani E, Villa D, Curiale S, Panzeri D, Badone FC, Landoni M. Isolation and characterization of a new mutant allele of *brachytic 2* maize gene. Molecular Breeding. 2007;**20**:83-91. DOI: 10.1007/s11032-006-9073-7

[61] Multani DS, Briggs SP, Chamberlin MA, Blakeslee JJ, Murphy AS, Johal GS. Loss of an MDR transporter in compact stalks of maize *br2* and sorghum *dw* mutants. Science. 2003;**302**:81-84. DOI: 10.1126/science.1086072

[62] Espinoza, V. J. y Vega, S. M. C. Maíces de alta frecuencia poliembriónica. In: Zavala, G. F.; Ortega, P. R.; Mejía, C. J. A.; Benítez, R. I. y Guillén, A. H. (Eds.). In: Memorias del XVIII Congreso Nacional de Fitogenética: Notas Científicas. SOMEFI. Chapingo, México. 2000

[63] Duarte FEVDO, Barros DDR, Girardi EA, Soares Filho WDS, Passos OS. Polyembryony and morphological seed traits in citrus rootstocks. Revista Brasileira de Fruticultura. 2013;**35**:246-254

[64] Frost HB. Seed reproduction: Development of gametes and embryos. The Citrus Industry. 1968;**2**:290-324

[65] Valdez LEL. Ganancia en calidad nutrimental del grano como respuesta asociada a la selección para poliembrionía en maíz [thesis]. Buenavista, Saltillo, Coahuila, México: Universidad Autónoma Agraria Antonio Narro; 2005

[66] Espinoza V. J.; Vega, S. M. C. y Jasso, C. D. Contenidos de grasa y proteína cruda en semillas de maíces poliembriónicos. In: Espinoza, V. J. y Del Bosque, C. J. (Eds.). In: Memoria del 2do. Taller Nacional de Especialidades de Maíz. Buenavista, Saltillo, Coahuila, México: Universidad Autónoma Agraria Antonio Narro; 1999. pp. 159-165

[67] Requena MC, Herrera RR, González CNA, Velázquez JE, Martínez MG, Cárdenas JDF. Alkaline cooking quality of polyembryonic and non-polyembryonic maize populations. Advance Journal of Food Science and Technology. 2011;**3**:259-268

[68] Braun A. Über Polyembryonie und Keimung von *Coelebogyne*. Ein Nachtrag zu der abhandlung über Parthenogenesis bei Pflanzen. Abh. Kon. Wiss; 1859. p. 109-263

[69] Lakshmanan KK, Ambegaokar KB. Polyembryony. In: Johri BM, editor. Embryology of Angiosperms. Berlin: Springer; 1984. p. 448-475. DOI: 10.1007/978-3-642-69302-1_9

[70] Spillane C, Steimer A, Grossniklaus U. Apomixis in agriculture: The quest for clonal seeds. Sexual Plant Reproduction. 2001;**14**:179-187. DOI: 10.1007/s00497-001-0117-1

[71] Bhojwani S, Bhatnagar S. The Embryology of Angiosperms. India: Vikas Publishing House Pvt Ltd; 1974. p. 355

[72] Johri BM, Ambegaokar KB. Embryology: Then and now. In: Johri BM, editor. Embryology of Angiosperms. Berlin: Springer; 1984. p. 1-52. DOI: 10.1007/978-3-642-69302-1

[73] Pérez BP, Márquez G. Origen de los brotes múltiples en la población de maíz IMM-UAAAN-BAP [thesis]. México, D.F: Universidad Autónoma de México; 2009

[74] Leroy JF. La polyembryonie chez les Citrus son ineret dans la cultura et amélioration. Revue internationale de botanique appliquée et d'agriculture tropicale. 1947;**27**:483-495. DOI: 10.3406/jatba.1947.6126

[75] Titova GE, Seldimirova OA, Kruglova NN, Galin IR, Batygina TB. Phenomenon of "Siamese embryos" in cereals in vivo and in vitro: Cleavage polyembryony and fasciations. Russian Journal of Developmental Biology. 2016;**47**:122-137. DOI: 10.1134/S10623604160300x61

[76] Andrade-Rodríguez M, Villegas-Monter A, Gutiérrez-Espinoza MA, Carrillo-Castañeda G, García-Velázquez A. Polyembryony and RAPD markers for identification of zygotic and nucellar seedlings in *Citrus*. Agrociencia. 2005;**39**:371-383

[77] Andrade-Rodríguez M, Villegas-Monter A, Carrillo-Castañeda G, García-Velázquez A. Polyembryony and identification of Volkamerian lemon zygotic and nucelar seedlings using RAPD. Pesquisa Agropecuária Brasileira. 2004;**39**:551-554. DOI: 10.1590/S0100-204X2004000600006

[78] Nakano M, Kigoshi K, Shimizu T, Endo T, Shimada T, Fujii H, Omura M. Characterization of genes associated with polyembryony and in vitro somatic embryogenesis in Citrus. Tree Genetics & Genomes. 2013;**9**:795-803. DOI: 10.1007/s11295-013-0598-8

[79] Paul P, Awasthi A, Kumar S, Verma SK, Prasad R, Dhaliwal HS. Development of multiple embryos in polyembryonic insertional mutant OsPE of rice. Plant Cell Reports. 2012;**31**:1779-1787. DOI: 10.1007/s00299-012-1291-3

[80] Kepiro JL, Roose ML. AFLP markers closely linked to a major gene essential for nucellar embryony (apomixis) in *Citrus maxima* × *Poncirus trifoliata*. Tree Genetics & Genomes. 2010;**6**:1-11. DOI: 10.1007/s11295-009-0223-z

[81] Gupta P, Shivanna KR, Mohan HY. Apomixis and polyembryony in the guggul plant, *Commiphora wightii*. Annals of Botany. 1996;**78**:67-72

[82] Mendes-Rodriguez C, Carmo-Oliveira R, Talavera S, Arista M, Ortiz PL, Oliveira PE. Polyembryony and apomixis in *Eriotheca pubescens* (*Malvaceae-Bombacoideae*). Plant Biology. 2005;**7**:533-540. DOI: 10.1055/s-2005-865852

[83] Espinoza Velázquez J, De León H. Apomixis, ¿un fenómeno en proceso de adopción natural en maíz? In: Valdez Reyna J, editor. Resultados de proyectos de investigación 2004. Buenavista, Saltillo, Coahuila, México: Dirección de investigación, Universidad Autónoma Agraria Antonio Narro; 2005. p. 245-252

[84] Leblanc O, Grimanelli D, Islam-Faridi N, Berthaud J, Savidan Y. Reproductive behavior in maize-tripsacum polyhaploid plants: Implications for the transfer of apomixis into maize. Journal of Heredity. 1996;**87**:108-111. DOI: 10.1093/oxfordjournals.jhered.a022964

[85] Leblanc O, Graminelli D, Hernandez-Rodriguez M, Galindo PA, Soriano Martinez AM, Perotti E. Seed development and inheritance studies in apomictic maize-*Tripsacum* hybrids reveal barriers for the transfer of apomixis into sexual crops. International Journal of Developmental Biology. 2009;**52**:585-596. DOI: 10.1387/ijdb.082813ol

[86] Stebbins GL. Apomixis in angiosperms. Botanical Review. 1941;**8**:507-542

[87] Khoklov SS. Evolutionary genetic problems of apomixis in angiosperms. In: Khoklov SS, editor. Apomixis and Plant Breeding. New Delhi: Amerind; 1976

[88] Kaur A, Ha CO, Jong K, Sands VE, Chan HT, Soepadno E, Ashton PS. Apomixis may be widespread among trees of the climax rain forest. Nature. 1978;**271**:440-442. DOI: 10.1038/271440a0

[89] Savidan Y, Carman JG, Dresselhaus T, editors. The Flowering of Apomixis: From Mechanisms to Genetic Enginneering. Mexico, D. F.: CIMMYT, IRD, European Commission DG VI (FAIR); 2001. p. 258. ISBN: 970-648-074-9

[90] Bulant C, Gallais A, Matthys-Rochon E, Prioul JL. Xenia effects in maize with normal endosperm: II. Kernel growth and enzyme activities during grain filling. Crop Science. 2000;**40**:182-189. DOI: 10.2135/cropsci2000.401182x

[91] Chassaigne A. Desarrollo de maíces especiales. In: S. Cabrera (ed.). IX Curso sobre producción de maíz. Araure, Venezuela; 2002. pp. 358-368

[92] Pilu R. The twin trait in maize. Maize Genetics Cooperation Newsletter. 2000;**74**:51

[93] Rodríguez HS, Castro GME. Estudio sobre herencia de semilla con dos embriones. In: Avances de investigación en maíz. Buenavista, Saltillo, Coahuila, México: Universidad Autónoma Agraria Antonio Narro; 1978. p. 19

[94] Kumar V, Malik SK, Pal D, Srinivasan R, Bhat SR. Comparative transcriptome analysis of ovules reveals stress related genes associated with nucellar polyembryony in citrus. Tree Genetics & Genomes. 2014;**10**:449-464. DOI: 10.1007/s11295-013-0690-0

[95] Puri A, Basha PO, Kumar M, Rajpurohit D, Randhawa GS, Kianian SF, Dhaliwal HS. The polyembryo gene (OsPE) in rice. Functional & Integrative Genomics. 2010;**10**:359-366. DOI: 10.1007/s10142-009-0139-6

[96] Espinoza Velazquez J, Valdez LL, González Vázquez VM, Musito Ramírez N, Gallegos JE, Sánchez LJ, Villarreal AG, Alcalá JM. Estudios genéticos sobre la poliembrionía en maíz. Análisis retrospectivo. In: Libro Científico Anual Agricultura, Ganadería y Ciencia Forestal UAAAN-2006. Buenavista, Coahuila, México: Universidad Autónoma Agraria Antonio Narro; 2008. pp. 2-8. ISBN: 978-968-844-059-9

[97] Musito Ramírez N, Espinoza Velázquez J, González Vázquez VM, Gallegos Solorzano JE, De León Castillo H. Seedling traits in maize families derived from a polyembryonyc population. Revista Fitotecnia Mexicana. 2008;**31**:399-402

Chapter 3

Molecular Breeding for Abiotic Stresses in Maize (*Zea mays* L.)

Asima Gazal, Zahoor Ahmed Dar and
Ajaz Ahmad Lone

Additional information is available at the end of the chapter

http://dx.doi.org/10.5772/intechopen.71081

Abstract

Abiotic constraints resulting from climate changes have widespread yield reducing effects on all field crops and therefore should receive high priority for crop breeding research. Conventional breeding has progressed a lot in building tolerant genotypes but abiotic stress tolerance breeding is limited by the complex nature of abiotic stress intensity, frequency, duration and timing, linkage drag of undesirable traits/genes with desirable traits; and transfer of favorable genes/alleles from diverse plant genetic resources limited by gene pool barriers giving molecular breeding a good option for breeding plant genotypes that can thrive in stress environments. Molecular breeding (MB) approaches viz., marker-assisted selection (MAS), marker-assisted backcrossing breeding (MABB), marker assisted recurrent selection (MARS) and genomic selection (GS) or genome wide selection (GWS) offer opportunities for plant breeders to develop high yielding maize cultivars with resilience to diseases in less time duration precisely. For complex traits (mainly abiotic stresses) where multiple QTLs control the expression, new strategies like marker assisted recurrent selection (MARS) and genomic selection (GS) are employed to increase precision and to reduce cost of phenotyping and time duration with disease resilience. This review discusses recent developments in molecular breeding for developing and improving abiotic stress resilience in field crops.

Keywords: cold, drought, waterlogging, climate change, salinity

1. Introduction

Even though climate change is one of the major current global concerns, it is not new. Several climate changes have occurred before, with dramatic consequences. Among them is the decrease in CO_2 content, 350 million years ago considered responsible for the leaf

appearance. It took nearly 40–50 million years for leaves to appear [1]. The massive volcanic eruptions were the second climatic change during the end-permian age in Siberia when lava erupted over 4 million km^3 onto the surface of earth [2] and today the volcanic eruption remnants cover an area of 5 million km^2. This volcanic eruption resulted in accumulation of organohalogens causing depletion of the ozone layer worldwide. Consequently, UV radiation burst was one of the cause of mass extinction resulting in wiping out 0.95 of all the species [2]. The end of the last ice age came to an end was the third major result of climatic changes causing long dry seasons. Hence, the annual plants survived dry seasons either as tubers or as dormant seeds leading to birth of agriculture in Fertile Crescent and then in other areas. The fourth climate change induced the Holocene flooding, ago which is now believed to be associated with collapsing of the ice sheets, resulting in rise of global sea level up to 1.4 m [3]. Rising sea levels caused massive migration towards the North Western areas which explained the domestication of plants and animals, which reached modern Greece, Balkans and Europe. During the last 5000 years, drought has historically been the main factor limiting crop production. Water availability has led to rise of multiple empires, while drought caused collapse of various civilizations viz., Mesopotamia, (6200 years ago), Yucatan Peninsula (1400 years ago), coastal Peru, (1700 years ago) and early bronze society in the south of Fertile Crescent [4, 5].

Climate changes have adverse impacts on food production, quality security [6]. The number of undernourished people would increase by 150% in the areas like, north of Africa and Middle East by year 2080 compared to 1990 and 300% in sub-Saharan Africa [7]. Agriculture is extremely vulnerable to climate change. Higher temperatures eventually reduce crop yields without discouraging weed, disease and pest challenges. Long-term production declines and short-term crop failures result from changes in precipitation patterns. Overall negative impact of climate change on agriculture is expected to threaten the global food security [8] which would probably increase unless early warning systems and breeding strategies are developed [9]. Climate change is reducing production while increasing hunger among populations. High temperatures with less precipitation over semi-arid regions would reduce yields of crops in the next two decades causing negative impacts on global food security and calorie consumption causing malnutrition [10]. Thus, agricultural productivity investments are needed to tackle the negative impacts of climate change on the health scenario and food security [8].

The most likely stresses within which plant breeding targets need establishing are: [11]

- High temperatures.
- Drought.
- Salinity.
- Biotic stresses.
- Increase in CO_2 concentration.

There is a three-fold relationship between climate change and agriculture. Firstly, agriculture contributes indirectly to climate change by emitting methane from rice fields, N_2O from fertilizers & manure and CO_2 emissions from field work, machinery, fertilizers and pesticides.

Second relation is the impact of these climate changes on agriculture caused by increased weather variability (extremes in temperature and precipitation), sea level rise and surge thus, inundating & ruining coastal agricultural lands, pathogen and pest pressures and decreased plant biodiversity. The third relation is that agriculture can itself become a potential moderator of climate change by mitigating climate change by carbon sequestration by having agroforestry, rotations with cover crops, green manure, conservation tillage, by changing inputs like going for organic farming, reducing fertilizers, using bio-fuels and by adapting to climate changes by breeding crop varieties with resilience to climate change by selective breeding and developing genetically modified organisms (GMOs) [12, 13].

To increase the efficiency of breeding pipelines, a combination of conventional, molecular, and transgenic breeding approaches will be needed. Breeding approaches are not mutually exclusive and are complimentary under most breeding schemes [14].

Plant breeders respond to climate related stresses in multiple ways:

- Selection and backcross breeding.
- Extensive managed stress screening experiments to develop superior tolerant germplasm via recurrent selection.
- Exploitation of alien genetic variation (Conserved Wild Relatives).
- Breeding for earliness and varieties with specific adaptation to specific ecologies.

One of the effective ways for crop production to grow or to stay stable under new challenges from climate change is through improved varieties developed by plant breeding. The genetic diversity of crop plants is the foundation for the sustainable development of new varieties for present and future challenges. For example, common beans biodiversity has been used by plant breeding to develop both heat and cold tolerant varieties grown from the hot Durango region in Mexico to the cold high altitudes of Colombia and Peru. Similar is the case with other crops too. Resource-poor farmers have been using genetic diversity intelligently over centuries to develop varieties adapted to their own environmental stress conditions.

Biotechnological tools: The tools of modern plant breeding include following:

- Molecular breeding (marker-assisted selection (MAS), marker-assisted backcrossing breeding (MABB), marker assisted recurrent selection (MARS), genome wide selection (GWS)).
- Genetic engineering.

1.1. Molecular breeding (MB)

The MB approach involves first identifying quantitative trait loci (QTLs) for tolerance to abiotic stresses. After identifying the markers associated with QTLs or genes for traits of interest, the candidate QTLs or genes can be introgressed in elite lines through marker-assisted backcrossing (MABC). Until recently, QTLs were identified by linkage mapping [15], but now association genetics has started to supplement these efforts in several crops [16, 17]. Nested association mapping, which combines the advantages of linkage analysis and association mapping in a single unified mapping population, is also being used for the genome-wide dissection of complex traits in maize [18]. Association mapping, compared with linkage mapping, is a high-resolution and relatively less expensive approach. In the near future, it is likely to be routinely used for identifying traits associated with abiotic stresses [16], particularly given the availability of high-throughput marker genotyping platforms [19]. An example of the systematic use of association mapping for drought tolerance is the collaborative project between Cornell University and CIMMYT (http://www.maizegenetics.net/drought-tolerance).

MABC helps in developing crops that are drought and heat tolerance, adapted to low rainfall and high temperature conditions. In rice, molecular breeding was used for one major effect QTL for submergence tolerance Sub1 QTL [20] and drought tolerance [21]. One of the difficulties of developing superior genotypes for abiotic stresses such as drought or heat is that these traits are generally controlled by small effect QTLs or several epistatic QTLs [22]. Incorporating QTLs by MABC has been limited, mainly because of the large sizes of the backcross populations. Therefore, marker-assisted recurrent selection (MARS) and genome wide selection (GWS) or genomic selection (GS): are used to overcome this problem of pyramiding several QTLs in the same genetic background [19, 23].

The estimated genetic gain by MARS or GWS is greater than obtained by using MABC for transferring QTLs /gene alleles for complex abiotic stress traits in one genetic background [24, 14]. The MARS approach is used routinely in private sector breeding programs [14, 25]. MABC and MARS require information on marker trait associations which is not necessarily required for GWS [26, 27]. GWS studies both phenotyping data as well as genome-wide marker profiling of a 'training population' and predictions of the genomic-estimated breeding values (GEBVs) of progeny GEBVs are calculated based on phenotyping and marker datasets. These values are used to select the superior progeny lines for advancement in breeding cycle [27, 28]. Several computational tools are available or are being developed to calculate GEBVs, such as the Best Linear Unbiased Prediction method and the geostatistical mixed model [29], (http://genomics.cimmyt.org/#Software).

2. Few case studies

2.1. Drought tolerance in rice

Birsa Vikas Dhan 111, an upland rice cultivar released in Jharkhand was bred by utilizing MABC for improved root growth QTLs towards improved performance under drought in a collaborative partnership programme between Birsa Agricultural University, Ranchi, Jharkhand, and CAZS-NR; Gramin Vikas Trust, Ranchi, Jharkhand. This variety is high yielding (out yielding recurrent parent by 10% in rainfed conditions) with good grain quality and matures early with tolerance towards. This specific QTL was identified by Adam Price in first instance (Aberdeen University, UK) and Brigitte Courtois (CIRAD, France/IRRI, Philippines). Here marker-assisted back-crossing breeding and marker assisted pyramid crossing was conducted to improve the morphological and root traits for drought tolerance of Indian rice variety, Kalinga III (*indica*) used as recurrent parent and Azucena, an upland japonica variety from Philippines as donor parent.

Five segments each from different chromosomes were targeted for introgression; four segments out of five carried the QTLs for root length and root thickness while as fifth segment had a recessive QTL for aroma. 24 NILs (Near isogenic lines) were evaluated in five field experiments in UAS Bangalore for root traits Dr. Shashidhar. The segment on chromosome number 9 with flanking markers viz., RM242-RM201 increased root length significantly both under drought & irrigated treatments thereby confirming the QTL from Azucena cultivar expressed well [21].

Significant number of QTLs associated with drought tolerance have been reported for drought tolerance. A QTL located on chromosome 9 has been found associated with spikelet fertility under drought stress and for root and shoot traits [30–32]. 'Teqing' a *indica* cultivar used as recurrent parent in a study with 'Lemont' as donor (*japonica*) several alleles from Lemont were found associated with improved drought tolerance [33]. Detection of qtl12.1 QTL for tolerance towards drought accounting for 51% of the genetic variance and located on chromosome 12 was reported by [34] localized to a 10.2-cM region (RM28048 and RM511).

NERICA rice varieties are promising for Africa. These varieties mature early and escape drought. Rice varieties hardier than NERICA are being, developed by maximizing the diversity of the African rice germplasm pool consisting of *Oryza glaberrima*, its wild relatives (*Oryza barthii*, *Oryza longistaminata*) and *Oryza sativa* landraces using both conventional breeding and biotechnology.

2.2. Drought tolerance in maize

One of the major limiting factors for maize production and productivity is inadequate soil moisture particularly during flowering and grain filling stages [35]. Studies on drought tolerance have focused on identifying the genetic basis of yield and its components and secondary traits viz., including anthesis-silking interval (ASI), root architecture and stay green. Stable genomic regions associated with flowering, maturity and yield components identified more than 1080 QTLs [36]. For narrow ASI, five QTLs were introgressed from a drought-tolerant donor Ac7643 through MABC to CML-247 an elite, drought-susceptible line. The selected lines out yielded the control under drought conditions while decreasing the yield advantage from mild to moderate drought stress [37].

In India several QTL mapping experiments on drought stress has been undertaken [38] and in China [39, 40]. In India, QTL mapping for maize drought tolerance identified major effect QTLs on chr. 1, 2, 8 and 10 after assessing a set of 230 CIMMYT developed RILs at Hyderabad and Karimnagar. A significant digenic epistatic QTL effect for kernel number ear^{-1} under drought stress was detected. A major QTL for ASI (anthesis-silking interval) and ear number per plant under drought stress was detected on chr. 1 (bin 1.03) and chr. 9 (bins 9.03–9.05) [39, 40] from a cross between X178 (tolerant line) and B73 which corresponded to several QTLs identified in different experiments carried on drought worldwide [41]. Several such identified 'consensus QTLs' would serve as good candidates in marker-assisted breeding to improve maize production under drought.

Drought resilient maize product pipeline:

Over 80% maize is grown as rain-fed crop, with avg. yield less than half of irrigated maize. Following are the few projects for developing drought resilient maize:

- Drought tolerant maize for Africa (DTMA),
- Water efficient maize for Africa (WEMA),
- Affordable accessible Asian drought tolerance maize project (AAA),
- Asian maize drought tolerance project (AMDROUT).

Drought tolerance maize varieties developed:

Variety	Trait + selection strategy	Developed By
ZM 309, 401, 423, 521, 623, 625 and 721	Conventional breeding	South Saharan Africa
KDV1, 4, 6	Conventional breeding	South Saharan Africa
WS103	Conventional breeding	South Saharan Africa
Melkassa 4	Conventional breeding	South Saharan Africa
WH 403, 502, 504, and ZMS402, 737	Conventional breeding	South Saharan Africa

2.3. Cold tolerance in rice

Tolerance of low temperature at both the vegetative and the reproductive stage is an important breeding objective for improving rice cultivars in the temperate and high altitude areas of the tropics and subtropics. Low temperatures during booting stage reduce yields by causing cold-induced male sterility. Cold prevents sugar accumulation in the pollen causing no starch build-up and hence no energy for pollen germination hampering grain production. Enzyme invertase regulated by hormone abscisic acid (ABA) transports sugar to tapetum before moving to the pollen and cold decreases the invertase levels in susceptible cultivars [42] lowering pollination and hence grain development. Several QTLs for cold tolerance were identified at booting stages on chromosomes 4 (Ctb1) and 8 (qCTB8) in Silewah (a javanica cultivar). Significant number of markers have been used by several workers [43, 44] to transfer cold tolerant gene (Ctb1) into japonica rice cultivars. Eight QTLs for booting-stage cold tolerance were identified in a RIL (recombinant inbred line) population derived from a cross between japonica and indica cultivars [45]. A QTL for cold induced wilting and necrosis tolerance has been fine mapped & identified on chromosome 12 [46, 47]. q*CTS4* fine mapped to 128-kb region on chromosome 4 associated with tolerance to stunning and yellowing of seedlings under cold contributed 40% of the phenotypic variation [48].

2.4. Salinity tolerance in rice

"White Leaf tip" is first symptom at vegetative stage in rice caused due to salinity stress followed by "Tip burning" which extends towards base. At reproductive stage papery sterile spikelets is another symptom resulting in huge losses and ultimately extreme high Salt Stress kills the rice plants. Central Soil Salinity Research Institute, Karnal is a pioneer institute in breeding for salinity resistant varieties. Few varieties developed by different approaches are as:

- Conventional:
  - **Pureline Selections** from local traditional cultivars **Pokkali, Nona Bokra and Kala-rata**:
  - **Damodar (CSR1), Dasal (CSR2), CSR3**.
  - **Pedigree: CSR10, 13, 23, 27, 30, 36.**
- Nonconventional:
  - Anther Culture: **CSR-21** for salinity.
  - **CSR: 28** for salinity and alkalinity.
- Other salt-tolerant rice varieties
  - Usar dhan 1, 2 & 3 (India);
  - BRRI dhan 40, BRRI dhan 41 (Bangladesh);
  - OM2717, OM2517, OM3242 (Vietnam).

MABC is being employed to efficiently transfer the Pokkali seedling stage salinity tolerant *Saltol* QTL into popular varieties such as IR64, BR11, BR28, Swarna, etc. *Saltol* QTL has been fine mapped on Chr. 1 shirt arm associated with the Na-K ratio (high K^+ & low Na^+ adsorptions) [49]. SKC1, a QTL for salt tolerance, maintains K^+ homeostasis in the tolerant cultivar and encodes HKT-type transporters [49]. QTLs for reproductive-stage salt tolerance are yet to be reported.

2.5. Submergence tolerance in rice

QTL *Sub1* fine mapped on chromosome 9 contributes 70% of the phenotypic variation for survival under submergence [50]. Two of the three ethylene-response factor (ERF) like genes induced by submergence were identified at this locus. [51] reported gene *Sub1A* gene responsible for submergence tolerance which has been integrated into Swarna by marker assisted backcross breeding [52] which demonstrated that QTLs controlling tolerance of abiotic stresses can be used to improve mega varieties in the target regions [53].

2.6. Waterlogging tolerance in maize

Over 18% of the total maize production area in South and Southeast Asia is frequently affected by floods and waterlogging problems, causing production losses of 25–30% annually [54]. Many QTLs for waterlogging tolerance at seedling stage have been reported [55]. A

F2:3 mapping population comprising 288 lines derived from HZ32 × K12 (sensitive) inbred lines studied under flooded and nonflooded conditions helped in identifying 25 and 34 QTLs accounting for between 4 and 37% of the genotypic variation to waterlogging tolerance. QTLs associated with plant height, root and shoot dry weight, total dry weight were identified in different experiments on chromosomes 4 and 9. In a F2 mapping population of B64 and teosinte (*Z. mays* ssp. Huehuetenangensis) QTLs associated with adventitious root formation under flooding were identified on chromosomes 3, 7 and 8 [56] confirming the potential use of teosinte as donor for waterlogging tolerance. A cross between *Z. mays* spp. Nicaraguensis (a different teosinte accession) and inbred line B73 helped in identifying QTLs for aerenchyma formation located on chromosomes 1, 5 and 8 [57]. These QTLs from different donors hence, provide a valuable genetic resource for breeding waterlogging tolerant maize.

2.7. Wheat drought and heat tolerance

Markers associated with a QTL for grain yield in wheat under drought has been identified at 4AL. 127 RILS were developed from a cross between Dharwar dry drought tolerant and Sitta drought susceptible [58]. XBE637912, Xwmc89, and Xwmc420 SSR markers were found linked to Grain Yield QTL.

3. Genetic engineering

Plant adaptation to environmental stresses is controlled by cascades of molecular networks. These activate stress responsive mechanisms to re-establish homeostasis and to protect and repair damaged proteins and membranes [59]. Abiotic stresses are multigenic, and hence difficult to control and engineer. Therefore, strategies like plant genetic engineering for building tolerance rely on gene expression involved in signaling pathways and regulatory pathways. Consequently, engineering genes that protect and maintain the function and structure of cellular components can enhance tolerance to stress [60].

4. Few case studies

4.1. Heat-tolerant basmati rice developed by over-expression of hsp101

Heat-tolerant basmati rice was developed by introducing *Arabidopsis thaliana hsp101 (Athsp101)* cDNA into the Pusa basmati 1 by *Agrobacterium* mediated transformation [61]. Transgenic lines were compared for survival after exposure to different levels of high-temperature stress {45°C for 3 h and then were placed at 28°C} with the untransformed control plants. It was reported that transgenic lines (15 and 43) survived heat stress as compared to the untransformed ones and the optimum temperature for rice growth throughout its life cycle is 25–31°C [61].

4.2. Barley gene in rice for drought tolerance

Barley gene *HVA7* was introduced into rice suspension cells using the Biolistic-mediated transformation method in rice for drought tolerance [62], *HVA7* is a late embryogenesis abundant (LEA) protein gene, from barley and this gene was regulated by the rice actin 1 gene promoter leading to high-level, constitutive accumulation of the HVA 7 protein in both leaves and roots of transgenic rice plants.

4.3. Yeast gene in tomato for salinity tolerance

In yeast (*Saccharomyces*) overexpression of *HAL 1* gene confers tolerance to salinity. So, introduction of this *HAL1* gene (using Plasmid pPM5 contained an *EcoRI: HindIII* fragment of 1.75 kb with the reinforced 35 S promoter, the *HAL1 ORF*, and the *nos* terminator) was done in Tomato (*Lycopersicon esculentum* cv P-73) [63]. Transgenic tomato (TG_3) was reported tolerant to salinity by maintaining K uptake in the presence of external Na.

4.4. Increased glycine betaine (GB) synthesis for salinity tolerance in cotton

Choline mono-oxygenase (CMO) is a major catalyst in glycine betaine (GB) synthesis. Glycine betaine is a osmolyte and overexpression of this osmolyte confers tolerance to salinity. This CMO gene cloned from *Atriplex hortensis* (AhCMO) was introduced into cotton (*Gossypium hirsutum* L.) via Agrobacterium mediation for development of Cotton plant having introduced CMO gene for glycine betaine (GB) [64].

4.5. Alteration in fatty acids: for cold stress tolerance

Plants such as squash and arabidopsis having high proportion of cis-unsaturated fatty acids are chilling resistant. Hence, the degree of unsaturation of fatty acids is closely related to chilling tolerance among the plants. Enzyme *glycerol-3-phosphate acetyl transferase* determines the phosphatidyl glycerol fatty acids unsaturation and hence cold tolerance.

5. Conclusions

Plant Genetic diversity and Plant Breeding are key elements in tackling climate change, and integration of plant breeding in climate change strategies is one of the best paths to sustainable food production by developing climate smart crops: Development of abiotic and biotic resistant crop varieties which cope with climatic vagaries, Varieties suited to new agricultural areas resulting due to shift in climatic pattern, Varieties with reduced total pesticide and fungicide consumption and hence, their reduced ill effects on environment which indirectly contribute to Climate Change. "It is not the strongest of the species who survive, nor the most intelligent, but the one most responsive to change." Let us be the difference we want to make to the world: Charles Darwin.

Author details

Asima Gazal[1]*, Zahoor Ahmed Dar[2] and Ajaz Ahmad Lone[2]

*Address all correspondence to: asimagazal@gmail.com

1 ITMU, ICAR-IIMR, New Delhi, India

2 DARS, Budgam (SKUAST-K), Srinagar, India

References

[1] Beerling DJ. The Emerald Planet: How Plants Changed Earth's History. Oxford, UK: Oxford University Press; 2007

[2] Beerling DJ, Osborne CP, Chaloner WG. Evolution of leaf-form in land plants linked to atmospheric CO_2 decline in the Late Palaeozoic era. Nature. 2001;**410**:352-354

[3] Turney CSM, Brown H. Catastrophic early Holocene sea level rise, human migration and the Neolithic transition in Europe. Quaternary Science Reviews. 2007;**26**:2036-2041

[4] Demenocal PB. Cultural responses to climate change during the late Holocene. Science. 2001;**292**:667-673

[5] Rosen AM. Environmental change at the end of early Bronze Age Palestine. In: De Miroschedji P, editor. L'urbanisation de la Palestine à l'âge du Bronze ancien. Oxford, UK: BAR International; 1990. p. 247-255

[6] Atkinson MD, Kettlewell PS, Poulton PR, Hollins PD. Grain quality in the Broadbalk wheat experiment and the winter North Atlantic oscillation. Journal of Agricultural Science, Cambridge. 2008;**146**:541-549

[7] Tubiello FN, Fischer G. Reducing climate change impacts on agriculture: global and regional effects of mitigation, 2000-2080. Technological Forecasting and Social Change. 2007;**74**:1030-1056

[8] Nelson GC, Rosegrant MW, Koo J, Robertson R, Sulser T, Zhu T, Ringler C, Msangi S, Palazzo A, Batka M, Magalhaes M, Valmonte-Santos R, Ewing M, Lee D. Climate Change: Impact on Agriculture and Costs of Adaptation. Food Policy Report. Washington, DC: International Food Policy Research Institute; 2009

[9] Brown ME, Funk CC. Food security under climate change. Science. 2008;**319**:580-581

[10] Lobell DB, Burke MB, Tebaldi C, Mastrandrea MD, Falcon WP, Naylor RL. Prioritizing climate change adaptation needs for food security in 2030. Science. 2008;**319**:607-610

[11] Ceccarelli S, Grando S, Maatougui M, Michael M, Slash M, Haghparast R, Rahmanian M, Taheri A, Al-yassin A, Benbelkacem A, Labdi M, Mimoun H, change NMC, paper a. plant breeding and climate changes. Journal of Agricultural Science. 2010;**148**:627-637

[12] Gomiero T, Paoletti MG, Pimentel D. Energy and Environmental Issues in Organic and Conventional Agriculture. Critical Reviews in Plant Sciences. 2008;**27**(4):239-254

[13] Niles JO, Brown S, Pretty J. Potential carbon mitigation and income in developing countries from changes in use and management of agricultural and forest lands. Philosophical Transactions. Royal Society of London. 2002;**360**:1621-1639

[14] Ribaut JM, de VMC, Delannay X. Molecular breeding in developing countries: Challenges and perspectives. Current Opinion in Plant Biology. 2010;**13**:213-218

[15] Varshney RK, Tuberosa R, editors. Genomics-Assisted Crop Improvement: Genomics Approaches and Platforms. Vol. I. The Netherlands: Springer; 2007

[16] Gupta PK, Rustgi S, Kulwal PL. Linkage disequilibrium and association studies in plants: Present status and future prospects. Plant Molecular Biology. 2005;**57**:461-485

[17] Hall D. Using association mapping to dissect the genetic basis of complex traits in plants. Briefings in Functional Genomics. 2010;**9**:157-165

[18] Yu J, Holland JB, McMullen MD, Buckler ES. Genetic design and statistical power of nested association mapping in maize. Genetics. 2008;**178**:539-551

[19] Varshney RK, Dubey A. Novel genomic tools and modern genetic and breeding approaches for crop improvement. Journal of Plant Biochemistry and Biotechnology. 2009;**18**:127-138

[20] Septiningsih EM, Pamplona AM, Sanchez DL, Neeraja CN, Vergara GV, Heuer S, Ismail AM, Mackill DJ. Development of submergence-tolerant rice cultivars: The Sub1 locus and beyond. Annals of Botany. 2009;**103**:151-160

[21] Steele KA, Price AH, Sashidhar HE, Witcombe JR. Marker-assisted selection to introgress rice QTLs controlling root traits into an Indian upland rice variety. Theoretical and Applied Genetics. 2006;**112**:208-221

[22] Messmer R, Fracheboud Y, Banziger M, Vargas M, Stamp P, Ribaut JM. Drought stress and tropical maize: QTL-by environment interactions and stability of QTLs across environments for yield components and secondary traits. Theoretical and Applied Genetics. 2009;**119**:913-930

[23] Tester M, Langridge P. Breeding technologies to increase crop production in a changing world. Science. 2010;**327**:818-822

[24] Bernardo R, Charcosset A. Usefulness of gene information in marker-assisted recurrent selection: A simulation appraisal. Crop Science. 2006;**46**:614-621

[25] Ribaut JM, Ragot M. Marker-assisted selection to improve drought adaptation in maize: The backcross approach, perspectives, limitations, and alternatives. Journal of Experimental Botany. 2006;**58**:351-360

[26] Jannink JL, Lorenz AJ, Iwata H. Genomic selection in plant breeding: From theory to practice. Briefings in Functional Genomics. 2010;**9**:166-177

[27] Heffner EL, Sorrells ME, Jannink JL. Genomic selection for crop improvement. Crop Science. 2009;**49**(1):12

[28] Meuwissen TH, Hayes BJ, Goddard ME. Prediction of total genetic value using genome-wide dense marker maps. Genetics. 2001;**157**:1819-1829

[29] Schulz-Streeck T, Piepho HP. Genome-wide selection by mixed model ridge regression and extensions based on geostatistical models. BMC Proceedings. 2010;**4**(Suppl. 1):S8

[30] Courtois B, McLaren G, Sinha PK, Prasad K, Yadav R, Shen L, Mapping QTL. associated with drought avoidance in upland rice. Molecular Breeding. 2000;**6**:55-66

[31] Li Z, Mu P, Li C, Zhang H, Li Z, Gao Y, Wang X. QTL mapping of root traits in a doubled haploid population from a cross between upland and lowland japonica rice in three environments. Theoretical and Applied Genetics. 2005;**110**:1244-1252

[32] Yue B, Xue WY, Xiong LZ, XQ Y, Luo LJ, Cui KH, Jin DM, Xing YZ, Zhang QF. Genetic basis of drought resistance at reproductive stage in rice: Separation of drought tolerance from drought avoidance. Genetics. 2006;**172**:1213-1228

[33] Xu JL, Lefftte HR, Gao YM, Fu BY, Torres R, Li ZK. QTLs for drought escape and tolerance identified in a set of random introgression lines of rice. Theoretical and Applied Genetics. 2005;**111**:1642-1650

[34] Bernier J, Kumar A, Ramaiah V, Spaner D Atlin G. A large effect QTL for grain yield under reproductive-stage drought stress in upland rice. Crop Science. 2007;**47**:507-518

[35] Joshi PK, Singh NP, Singh NN, Gerpacio RV, Pingali PL. Maize in India: Production Systems, Constraints, and Research Priorities. Mexico, D.F.: CIMMYT; 2005

[36] Ribaut JM, Betran J, Monneveux P, Setter T. Drought tolerance in maize. In: Bennetzen JL, Hake SC, editors. Handbook of Maize. New York: Springer; 2009. p. 311-344

[37] Ribaut JM, Ragot M. Marker-assisted selection to improve drought adaptation in maize: The backcross approach, perspectives, limitations, and alternatives. Journal of Experimental Botany. 2007;**58**:351-360

[38] Prasanna BM, Beiki AH, Sekhar JC, Srinivas A, Ribaut JM. Mapping QTLs for component traits influencing drought stress tolerance of maize in India. Journal of Plant Biochemistry and Biotechnology. 2009;**18**:151-160

[39] Xiao YN, Li XH, George ML. Quantitative trait loci analysis of drought tolerance and yield in maize in China. Plant Molecular Biology Reporter. 2005;**23**:155-165

[40] Hao Z, Li X, Xie C. Two consensus quantitative trait loci clusters controlling anthesis-silking interval, ear setting and grain yield might be related with drought tolerance in maize. The Annals of Applied Biology. 2008;**153**:73-83

[41] Tuberosa R, Salvi S, Giuliani S. Genome-wide approaches to investigate and improve maize response to drought. Crop Science. 2007;**47**:S120-S141

[42] Oliver SN, Dennis ES, Dolferus R. ABA regulates apoplastic sugar transport and is a potential signal for coldinduced pollen sterility in rice. Plant and Cell Physiology. 2007;**48**:1319-1330

[43] Saito K, Hayano-Saito Y, Maruyama-Funatsuki W, Sato Y, Kato A. Physical mapping and putative candidate gene identification of a quantitative trait locus Ctb1 for cold tolerance at booting stage of rice. Theoretical and Applied Genetics. 2004;**109**:515-522

[44] Kuroki M, Saito K, Matsuba S, Yokogami N, Shimizu H, Ando I, Sato Y. A quantitative trait locus for cold tolerance at the booting stage on rice chromosome 8. Theoretical and Applied Genetics. 2007;**115**:593-600

[45] Andaya VC, Mackill DJ. QTLs conferring cold tolerance at the booting stage of rice using recombinant inbred lines from a japonica x indica cross. Theoretical and Applied Genetics. 2003;**106**:1084-1090

[46] Andaya VC, Mackill DJ. Mapping of QTLs associated with cold tolerance during the vegetative stage in rice. Journal of Experimental Botany. 2003;**54**:2579-2585

[47] Andaya VC, Tai TH. Fine mapping of qCTS12 locus, a major QTL for seedling cold tolerance in rice. Theoretical and Applied Genetics. 2006;**113**:467-475

[48] Andaya VC, Tai TH. Fine mapping of the qCTS4 locus associated with seedling cold tolerance in rice (*Oryza sativa* L.). Molecular Breeding. 2007;**20**:349-358

[49] Ren Z, Gao J, Li L, Cai X, Huang W, Chao D, Zhu M, Wang Z, Luan S, Lin H. A rice quantitative trait locus for salt tolerance encodes a sodium transporter. Nature Genetics. 2005;**37**:1141-1146

[50] Xu K, Xu X, Ronald PC, Mackill DJ. A high-resolution linkage map in the vicinity of the rice submergence tolerance locus Sub1. Molecular & General Genetics. 2000;**263**:681-689

[51] Xu K, Xia X, Fukao T, Canlas P, Maghirang-Rodriguez R, Heuer S, Ismail AI, Bailey-Serres J, Ronald PC, Mackill DJ. Sub1A is an ethylene response factor-like gene that confers submergence tolerance to rice. Nature. 2006;**442**:705-708

[52] Neeraja CN, Maghirang-Rodriguez R, Pamplona A, Heuer S, Collard BCY, Septiningsih EM, Vergara G, Sanchez D, Xu K, Ismail AM, Mackill DJ. A marker-assisted backcross approach for developing submergence-tolerant rice cultivars. Theoretical and Applied Genetics. 2007;**115**:767-776

[53] Mackill DJ. Breeding for resistance to abiotic stresses in rice: The value of quantitative trait loci. In: Lamkey KR, Lee M, editors. Plant breeding: The Arnel R. Hallauer International Symposium. Ames, IA: Blackwell; 2006. p. 201-212

[54] Zaidi PH, Maniselvan P, Srivastava A, Yadav P, Singh RP. Genetic analysis of waterlogging tolerance in tropical maize (*Zea mays* L.). Maydica. 2010;**55**:17-26

[55] Qiu F, Zheng Y, Zhang Z, Xu S. Mapping of QTL associated with waterlogging tolerance during the seedling stage in maize. Annals of Botany. 2007;**99**:1067-1081

[56] Mano Y, Omori F, Muraki M, Takamizo T. QTL mapping of adventitious root formation under flooding conditions in tropical maize. Breeding Science. 2005;**55**:343-347

[57] Mano Y, Omori F, Loaisiga CH, Bird RM. QTL mapping of aboveground adventitious roots during flooding in maize x teosinte *Zea nicaraguensis* backcross population. Plant Roots. 2009;**3**:3-9

[58] Kirigwi FM, Van Ginkel M, Brown-Guedira G, Gill BS, Paulsen GM, Fritz AK. Markers associated with a QTL for grain yield in wheat under drought. Molecular Breeding. 2007;**20**:401-413

[59] Wang W, Vinocur B, Altman A. Plant responses to drought, salinity and extreme temperatures: Towards genetic engineering for stress tolerance. Planta. 2003;**218**:1-14

[60] Vinocur B, Altman A. Recent advances in engineering plant tolerance to abiotic stress: Achievements and limitations. Current Opinion in Biotechnology. 2005;**16**:123-132

[61] Agarwal SK, Agarwal M, Grover A. Heat-tolerant basmati rice engineered by overexpression of hsp101. Plant Molecular Biology. 2003;**51**(5):677-686

[62] Xu D, Duan X, Wang B, Hong B, Ho TD, Wu R. Expression of a late embryogenesis abundant protein gene, HVA7, from barley confers tolerance to water deficit and salt stress in transgenic rice. Plant Physiology. 1996;**110**:249-257

[63] Carmina G, Rus AM, Bolarin MC, Lopez-Coronado JM, Montesinos C, Serrano R, Moreno V. The yeast *HAL1* gene improves salt tolerance of transgenic tomato. Plant Physiology. 2000;**123**(1):393-402

[64] Zhang H, Dong H, Li W, Sun Y, Chen S, Kong X. Increased glycine betaine synthesis and salinity tolerance in AhCMO transgenic cotton lines. Molecular Breeding. 2009;**23**:289-298

Chapter 4

Genetic Variability for Resistance to Leaf Blight and Diversity among Selected Maize Inbred Lines

Abera Wende, Hussein Shimelis and
Eastonce T. Gwata

Additional information is available at the end of the chapter

http://dx.doi.org/10.5772/intechopen.70553

Abstract

Maize (*Zea mays* L.) is an important staple food crop in sub-Saharan Africa (SSA). The productivity of the crop is limited partly by the leaf blight disease caused by *Exserohilum turcicum*. In breeding for resistance to leaf blight, the germplasm needs to be well-characterized in order to design efficient breeding programs. This study evaluated the (i) genetic variability among maize inbred lines and (ii) diversity of selected medium to late maturity tropical maize inbred lines for hybrid breeding. Plants of 50 maize inbred lines were artificially inoculated in the field during 2011 and 2012. Disease severity and incidence as well as grain yield were measured. A subset of 20 elite maize inbred lines was genotyped using 20 SSR markers. The germplasm showed significant differences in reaction to leaf blight and were classified as either resistant or intermediate or susceptible. Mean disease severity varied from 2.04 to 3.25. Seven inbred lines were identified as potential sources of resistance to leaf blight for the genetic improvement of maize. The genotyping detected 108 alleles and grouped the inbred lines into five clusters consistent with their pedigrees. The genetic grouping in the source population will be useful in the exploitation of tropical maize breeding programs.

Keywords: leaf blight, inbred line, mid-altitude, maize, pedigree

1. Introduction

Maize (*Zea mays* L.) is an important staple food crop in sub-Saharan Africa (SSA). It is the third most important cereal crop after wheat and rice [1]. It is used for both livestock feeds and human consumption. In SSA, maize accounts for about 70% of the human food [2]. The demand for maize is expected to increase by >90.0% in SSA by 2020 [3]. However, the productivity of the crop is limited by several abiotic and biotic stresses. Among these abiotic factors,

insect pests, such as the stem borers and weevils, cause considerable economic damage on the crop [4, 5]. In addition, fungal diseases such as gray leaf spot (*Cercospora zeae-maydis* Tehon & Daniels), common leaf rust (*Puccinia sorghi* Schr.), and turcicum leaf blight (TLB) (*Exserohilum turcicum*) often pose a serious threat to maize production [6].

In particular, TLB, also known as the northern corn leaf blight, can devastate the crop in high rainfall, humid areas [6, 7]. TLB reduces the seed quality, resulting in diminished germination capacity, low sugar content as well as predisposition to stalk rot [8, 9]. The use of resistant varieties is an inexpensive method for combating TLB [10]. Currently, there are efforts to incorporate durable resistance into maize germplasm particularly in SSA where some commercial varieties as well as elite parental inbred lines are reportedly vulnerable to TLB [11, 12]. For example, in Ethiopia, maize productivity is low (averaging about 2.5 t/ha) in the smallholder production systems partly due to TLB and other stresses. Spurred by the need to enhance maize productivity for farmers, the national maize improvement program in Ethiopia recently embarked on a breeding project aimed at developing leaf blight resistant hybrid varieties that are adapted to the major maize-growing areas of the country which are predominantly in the mid-altitude to subhumid agroecologies [13]. However, hybrid breeding for resistance to leaf blight requires knowledge of the genetic variability of the germplasm in terms of its reaction to TLB as well as its characterization into distinct genetic groups that can be hybridized in order to exploit heterosis.

The variability in the host (maize) plant resistance to the disease occurs in either the qualitative or the quantitative form. The qualitative form of resistance is race specific and is governed by a single or few genes but the quantitative form of resistance is race nonspecific and polygenic [14, 15]. In addition, qualitative resistance can break down due to the emergence of new virulent races of the pathogen through genetic mutation and recombination events [12, 15]. The pathogen *E. turcicum* exhibits a wide range of variability [16], and new races are capable of overcoming previously resistant varieties [7]. For instance, the resistance conferred by the Htn gene(s) is characterized by chlorotic and necrotic lesions or lesions surrounded by a yellow-to-light-brown margin (without spore formation), which limits the growth and spread of the disease [12, 14]. In contrast, the resistance conferred by Htn gene is expressed as a delay in lesion formation typically showing at the pollination stage [17, 18]. Lesion size, together with area under disease progress curve (AUDPC) as well as disease severity and incidence, are commonly used in evaluating maize genotypes for resistance to TLB [19, 20]. However, phenotypic evaluations in conventional breeding approaches are unable to detect the presence of favorable alleles in the germplasm. Therefore, marker-assisted selection and DNA fingerprinting techniques have been effectively used to increase the efficiency of conventional breeding, particularly the time required for developing new improved varieties in maize [12].

The presence of discrete genetic groups among inbred lines is attributed to increased allelic diversity which is useful in optimizing hybrid vigor. Assigning inbred lines into well-differentiated genetic clusters can reduce the creation and evaluation of many undesirable crosses [21]. Molecular markers assist in characterizing inbred lines and in establishing distinct clusters of genotypes based on genetic diversity, which is useful in maize breeding programs [22, 23]. Molecular markers were applied successfully to allocate maize germplasm into heterotic groups [24–26]. In a study which compared different markers for their effectiveness in estimating genetic grouping among maize inbred lines, SSR markers revealed the highest level of polymorphism due to their codominant nature and high number of alleles per locus [27]. Therefore, the study reported in this chapter was designed to evaluate the (i) genetic variability in reaction to TLB among maize inbred lines under field conditions and (ii) diversity of selected medium to late maturity tropical maize inbred lines for hybrid breeding using selected SSR markers.

2. Materials and methods

2.1. Field evaluation

2.1.1. Germplasm and testing location

Fifty inbred lines were used in the study. The lines were adapted to the mid-altitude agroecologies in Ethiopia and were obtained from the national maize research program and the international maize and wheat improvement center (CIMMYT). Inbred line CML-197, which was obtained from CIMMYT, served as susceptible check (**Table 1**). The field trial was conducted at Bako (37°09′ E; 09°06′ N; 1650 m above sea level). It receives approximately 1200 mm rainfall annually (**Table 2**) and is representative of the mid-altitude subhumid agroecological region in Ethiopia.

2.1.2. Field experiments

Inbred lines were evaluated using the lattice design with three replications. Trials were conducted for two consecutive seasons (in 2011 and 2012) during the main rainy season (May to September) in Ethiopia. The seed of each genotype was planted manually in the field in a two-row plot 5.1 m long × 0.75 m at 30.0 cm intra-row spacing. Phosphorus (in the form of diammonium phosphate) was applied once at planting at 100.0 kg/ha. Nitrogen fertilizer (in the form of urea) was applied at 100.0 kg/ha in two splits with 50% at planting and the rest at 37 days after emergence. Standard maize trial management practices were applied throughout each season at the location.

2.1.3. Leaf blight inoculum collection, preparation and inoculation

Isolates of *E. turcicum* were obtained from diseased maize leaf samples that were collected from fields where the disease is prevalent. The infected leaves were excised into small sections (approx. 1.0 cm^2 each) prior to surface sterilization using 2.5% Sodium hypochlorite for about 3 min and subsequently rinsed with sterile distilled water and blot-dried before plating on PDA in petri dishes for incubation at room temperature for 3–4 days. Pure cultures were prepared by subculturing from the isolation plates followed by incubation for 7–10 days in order to obtain sufficient growth. The inoculum was prepared by flooding the cultures with sterile distilled water and scrapping the surface with microscopic slides to dislodge the conidia and then filtered using cheese cloth after which the concentration of the conidia suspension was adjusted to approximately 105 conidia per milliliter using a hemocytometer [28].

Entry	Pedigree	Origin
1	CML 202	CIMMYT
2	CML 442	CIMMYT
3	CML 312	CIMMYT
4	CML 464	CIMMYT
5	Gibe-1-91-1-1-1-1	BAKO
6	CML 445	CIMMYT
7	CML 443	CIMMYT
8	CML 197	CIMMYT
9	A-7033	BAKO
10	CML 205/208//202-X-2-1-2-B-B-B	BAKO
11	CML 395	CIMMYT
12	F-7215	BAKO
13	DE-78-Z-126-3-5-5-1-1	BAKO
14	30H83-7-1-1-1-2-1	BAKO
15	I100E-1-9-1-1-1-1-1	BAKO
16	SZYNA99F2-81-4-3-1	BAKO
17	X1264DW-1-2-1-1-1	BAKO
18	124-b (113)	BAKO
19	SC22	BAKO
20	SC715-121-1-3	BAKO

Table 1. The pedigree and origin of maize inbred lines that were evaluated for diversity using SSR markers.

Maize plants growing in the field were inoculated at the four to six leaf growth stages during the middle of the main rainy season (mid-July) in Ethiopia. The inoculations were accomplished by spraying (manually, with the aid of an atomizer) the maize plant with the conidia suspension until runoff after which fine mist water was sprayed over the inoculated plants in order to create conducive conditions for disease development. This inoculation procedure was carried out during the evening when there was sufficient moisture in the air.

2.1.4. Data collection and analysis

In each season, the disease was visually assessed in the field 2–3 weeks after inoculation. Ten randomly selected plants were tagged and used for successive disease assessments. Plants were rated at 10-day intervals for percent incidence, lesion length, and lesion width. In order to determine the rate of lesion expansion, 2 lesions out of the 10 plants were measured (and marked for subsequent tracing) at 10-day intervals.

Month	2011			2012		
	Rainfall (mm)	Temperature (C⁰)	RH (%)	Rainfall (mm)	Temperature (C⁰)	RH (%)
January	15.90	20.20	58.00	0.00	20.40	52.70
February	2.00	20.90	50.90	4.40	21.80	47.50
March	58.80	21.90	53.90	16.20	23.00	48.90
April	68.10	20.40	52.40	30.70	24.00	62.50
May	222.20	21.30	58.50	92.8	23.00	55.60
June	295.00	19.90	67.50	153.30	20.20	66.90
July	224.10	19.30	69.30	138.20	19.50	76.00
August	294.60	19.10	75.60	263.60	19.70	64.00
September	131.30	20.00	65.90	157.50	20.10	74.40
October	53.20	20.20	59.80	6.00	21.00	50.50
November	60.10	20.00	59.80	17.10	20.30	49.70
December	0.00	19.80	54.50	6.70	21.5	45.70
Total	1425.30			886.50		

RH = relative humidity.

Table 2. Average monthly rainfall, temperature, and relative humidity at Bako during the 2011 and 2012 cropping seasons.

Disease severity was scored using a scale of 1–5 where:

1.0 = very slightly infected, one or two restricted lesions on lower leaves or trace.

2.0 = slight-to-moderate infection on lower leaves, a few scattered lesions on lower leaves.

3.0 = abundant lesions on lower leaves, a few on middle leaves.

4.0 = abundant lesions on lower and middle leaves extending to upper leaves.

5.0 = abundant lesions on all leaves, plant may be prematurely killed by blight.

The AUDPC was determined from the disease severity scores obtained in both seasons. The AUDPC parameter was calculated using Eq. (1) below as described previously [29]:

$$\text{AUDPC} = \sum_{i=1}^{n-1} \frac{(y_i + y_{i+1})(t_{i+1} - t_i)}{2} \quad (1)$$

where n = number of observations, t_i = number of days after planting for the ith disease assessment, and y_i = disease severity.

The parameter was used to quantify the epidemic from the beginning to the peak of the disease. The grain yield was calculated using the average shelling percentage of 80% adjusted to 12.5% moisture. Data sets of the quantitative measurements from individual trials were subjected to standard analysis of variance procedures using the GenStat release 14.2 computer software program [30].

2.2. Marker evaluation

2.2.1. Germplasm

Twenty maize inbred lines were used in the study. Eight of these inbred lines were originally developed for the mid-altitude and subhumid agroecologies at CIMMYT, whereas the remainder was developed by the local Ethiopian maize research program and was well adapted to mid-altitude areas. The local inbred lines were developed from three heterotic groups (that are commonly used in the country) namely Kitale synthetic II, Ecuador 573, and Pool 9A.

2.2.2. DNA sampling

DNA was collected from 3- to 4-week-old plants (tagged for identification), using Whatman FTA cards and the modified protocol of FTA paper technology [31]. Ten DNA samples from each of the 20 inbred lines were then bulked (in order to eliminate variation within each entry) and used for the diversity analysis at the INCOTEC-PROTEIOS laboratory in South Africa (Incotec, SA Pty. Ltd., South Africa) utilizing 20 SSR markers. PCR products of all of the 20 primers were fluorescently labeled and separated by capillary electrophoresis on an ABI 3130 automatic sequencer (Applied Biosystems, Johannesburg, South Africa). Analysis was performed using GeneMapper 4.1. The data matrices of the genetic distances were used to create the dendrogram using the unweighted pair group method with arithmetic mean allocated (UPGMA). The polymorphism information content (PIC) was calculated as:

$PIC = 1 - \sum fi$.

where fi is the frequency of the ith allele [32].

3. Results and discussion

3.1. Disease development and severity

Disease ratings were significantly different among the 50 inbred lines ($P < 0.001$), and 11 were classified as resistant, 26 as intermediate, whereas the remainder was classified as susceptible (**Tables 3** and **4**). The resistant inbred lines (e.g., 136-a and 142-1-e) attained lower disease severity scores compared to the susceptible check CML-197 (**Tables 3** and **4**). No accession was immune to the disease. In addition, there were highly significant ($P < 0.001$) differences for lesion length among inbred lines in both 2011 and 2012. The inbred lines Pool9A-4-4-1-1-1, SZSYNA-99-F2-803-4-1, and CML 197 showed comparatively larger lesion lengths, whereas the lesion length of CML 202 and CML 312 showed consistently small lesion lengths over the two seasons. Resistance to *E. turcicum* in maize germplasm was previously associated with a reduction in percent leaf area as well as small lesions [33].

The significant differences detected among genotypes in this study across the 2 years (cropping seasons) was attributable to a range of factors such as favorable climatic conditions, the inoculation method employed, and proper disease rating. In other studies, the development of NLB was attributed to pathogenic fitness and environmental conditions [34]. In Ethiopia,

http://dx.doi.org/10.5772/intechopen.70553

No.	Inbred line	DSS	Reaction type	Incidence (%)	Lesion length (cm)	AUDPC	TSW	Yield (t ha^{-1})
1	CML 202	2.00	R	46.81	9.88	408.3	223.3	2.22
2	CML442	2.734	I	78.43	13.40	612.5	223.3	2.40
3	CML 312	2.413	I	61.52	10.35	385.0	276.7	3.03
4	CML 464	2.210	I	55.64	13.82	595.0	223.3	3.79
5	Gibe-1-91-1-1-1-1	2.534	I	71.32	14.57	408.3	321.7	2.90
6	CML 445	2.523	I	65.20	14.02	571.7	213.3	3.34
7	CML 443	2.934	S	69.61	13.48	595.0	211.7	2.07
8	Gibe-1-158-1-1-1-1	2.496	I	66.42	11.37	507.5	281.7	3.43
9	A7033	2.881	S	68.63	15.37	641.7	273.3	2.58
10	(CML 205/CML208//CML 202)-X2-1-2-B-B-B	2.696	S	83.58	15.88	571.7	300.0	5.60
11	CML395	2.388	I	71.08	14.07	420.0	338.3	4.96
12	CML 444	2.526	I	69.12	18.28	443.3	260.0	2.95
13	DE-78-Z-126-3-2-2-1-1	2.688	S	67.89	14.48	536.7	280.0	4.14
14	30H83-7-1-1-1-2-1	2.00	R	53.19	10.90	495.8	210.0	3.14
15	ILoo'E-1-9-1-1-1-1-1	2.00	R	56.62	15.62	420.0	346.7	4.83
16	SZSYNA-99-F2-814-3-1	2.00	R	42.40	10.77	466.7	315.0	2.46
17	X1264DW-1-2-1-1-1-1	2.889	S	70.59	15.00	571.7	213.3	1.94
18	124-b(113)	2.559	I	59.80	15.27	606.7	365.0	3.53
19	SC22	2.760	S	85.78	14.72	501.7	271.7	3.56
20	SC-715-1211-3	2.466	I	67.40	13.47	396.7	336.7	3.45
21	DE-105-Z-126-30-1-2-2-1	2.00	R	61.27	14.55	420.0	235.0	2.89
22	Gibe-1-20-2-2-1-1	2.663	S	69.12	18.78	501.7	301.7	2.62
23	Kuleni-0080-4-2-1-1-1-1	2.022	I	61.52	16.38	449.2	326.7	3.72
24	Pool9A-4-4-1-1-1	2.677	S	68.63	21.35	670.8	288.3	4.85
25	30H83-5-1-4-2-1-1	2.486	I	63.97	16.27	484.2	308.3	4.27
26	Iloo'E-5-5-3-1	2.639	I	74.26	13.48	560.0	328.3	4.41
27	SZSYNA-99-F2-2-7-3-1-1	2.00	R	57.35	11.77	478.3	206.7	2.77
28	SC-715-154-1-1	2.206	I	65.20	11.97	402.5	280.0	5.89
29	BH6609(F2)-10-2-1-2-1	2.333	I	61.76	11.83	402.5	300.0	3.98
30	143-5-I	2.305	I	60.29	15.48	420.0	325.0	6.84
31	144-7-b	1.90	R	58.09	12.87	385.0	330.0	4.45

No.	Inbred line	DSS	Reaction type	Incidence (%)	Lesion length (cm)	AUDPC	TSW	Yield (t ha^{-1})
32	(LZ-955459/LZ955357)-B-1-B-B	2.369	I	67.16	12.20	431.7	256.7	2.98
33	139-5-j	2.00	R	53.43	13.78	385.0	258.3	2.56
34	30H83-56-1-1-1-1-1	2.351	I	57.35	10.22	495.8	205.0	3.57
35	SZSYNA-99-F2-80-3-4-1	2.653	I	73.53	20.05	525.0	293.3	3.15
36	124-b(109)	2.901	S	81.86	15.48	536.7	310.0	5.54
37	F7215	2.417	I	63.73	14.72	455.0	393.3	3.86
38	136-a	1.80	R	51.47	13.82	238.0	396.7	4.41
39	DE-78-Z-126-3-2-1-2-1	2.631	I	70.83	14.85	595.0	286.7	3.83
40	Gibe-1-186-2-2-1	2.549	I	51.96	14.88	350.0	373.3	2.70
41	Pool9A-128-5-1-1-1	2.718	I	71.43	13.12	595.0	278.3	2.45
42	30H83-7-1-5-1-1-1-1	2.00	R	52.45	12.05	379.2	220.0	2.60
43	SZSYNA-99-F2-3-6-2-1	2.587	I	70.83	12.33	618.3	256.7	2.36
44	SC-715-13-2-1	2.434	I	61.76	12.87	420.0	248.3	2.34
45	SC-22-430(63)	3.033	S	80.15	11.57	478.3	311.7	2.48
46	Kuleni-C1-101-1-1-1	3.028	S	75.49	17.07	700.0	258.3	2.84
47	Iloo'E-1-12-4-1-1	2.355	I	51.96	10.30	443.3	276.7	2.43
48	(DRB-F2-60-1-2)-B-1-B-B-B	2.791	S	75.98	16.23	600.8	270.0	2.66
49	142-1-e	2.00	R	62.01	15.02	595.0	323.3	3.94
50	CML 197	3.028	S	88.48	18.07	525.0	271.7	50
	LSD	0.4260	—	18.513	7.504	129.93	72.64	1.465
	Pr > f	**	—	**	**	**	**	**
	CV (%)	3.3	—	17.6	10.6	16.2	15.9	25
	Overall mean	2.486	—	65.49	14.16	493.9	284.1	3.52

DSS = disease severity score (0.00–5.00); R = resistant (1.0–2.00); I = intermediate (2.10–2.50); susceptible (2.51–5.00); and TSW = thousand seed weight.

** = Significant at 0.05 and 0.01 probability levels, respectively.

Table 3. Maize leaf blight reactions, grain yield, and thousand seed weight of 50 inbred lines tested during 2011 at Bako research Center in Ethiopia.

the disease infection and epidemics in maize occur largely during the main production season particularly in the wet and humid areas. Therefore, breeding for resistance to the disease in such areas is critical.

Disease severity scores in both cropping seasons were significantly different ($P < 0.01$) (**Tables 3** and **4**). During the two seasons, the lowest severity scores were observed for the inbred lines CML 202, 144-7-b, and 142-1-e. In contrast, relatively high severity scores were

No.	Inbred line	DSS	Reaction type	Incidence (%)	Lesion length (cm)	Lesion width (cm)	TSW	Yield (t/ha)
1	CML 202	2.39	R	40.69	12.00	1.33	173	2.15
2	CML442	2.69	S	72.55	13.67	1.67	210	2.67
3	CML 312	2.47	I	64.22	12.33	0.83	220	3.25
4	CML 464	1.92	R	52.45	13.00	1.03	207	3.01
5	Gibe-1-91-1-1-1-1	2.53	S	74.02	20.33	1.50	260	3.76
6	CML 445	2.42	I	65.69	14.33	1.17	207	3.36
7	CML 443	2.97	S	64.71	13.00	1.00	183	1.93
8	Gibe-1-158-1-1-1-1	2.39	I	58.82	12.00	1.57	270	2.93
9	A7033	2.81	S	58.82	13.33	1.33	240	2.41
10	(CML 205/CML208//CML 202) -X2-1-2-B-B-B	2.64	S	86.76	22.33	1.83	237	5.83
11	CML395	2.33	I	70.59	21.67	2.00	313	5.04
12	CML 444	2.61	S	65.69	23.33	2.00	230	2.67
13	DE-78-Z-126-3-2-2-1-1	2.67	S	65.20	18.33	1.50	250	2.6
14	30H83-71-1-1-2-1	1.89	R	39.71	13.33	1.67	187	2.91
15	ILoo'E-1-9-1-1-1-1-1	2.14	I	54.41	23.67	1.33	293	4.62
16	SZSYNA-99-F2-814-3-1	1.69	R	27.94	14.00	1.00	257	2.01
17	X1264DW-1-2-1-1-1-1	2.36	I	72.55	19.33	1.33	183	1.92
18	124-b(113)	2.34	I	45.10	16.33	1.67	303	3.13
19	SC22	2.05	I	91.18	16.67	2.00	230	3.14
20	SC-715-121-1-3	3.07	S	70.10	16.00	2.17	270	2.64
21	DE-105-Z-126-30-1-2-2-1	1.57	R	69.61	20.67	1.83	230	3.42
22	Gibe-1-20-2-2-1-1	2.48	I	77.45	25.00	1.33	287	3.08
23	Kuleni-0080-4-2-1-1-1-1	4.29	I	58.33	20.33	1.33	283	3.8
24	Pool9A-4-4-1-1-1	2.42	I	62.25	25.67	1.67	270	5.35
25	30H83-51-4-2-1-1	2.47	I	67.16	22.00	2.00	260	4.39
26	Iloo'E-5-5-3-1	2.61	S	77.94	14.33	1.00	260	3.5
27	SZSYNA-99-F2-2-7-3-1-1	2.22	I	55.88	15.00	1.50	170	2.88
28	SC-715-154-1-1	2.14	I	73.53	15.67	1.83	217	5.01
29	BH6609(F2)-10-2-1-2-1	2.33	I	54.90	15.33	1.53	243	1.74
30	143-5-I	2.08	I	51.96	18.00	2.17	273	5.95
31	144-7-b	1.89	R	59.31	18.00	1.00	333	3.47
32	(LZ-955459/LZ955357)-B-1-B-B	2.28	I	67.65	16.67	1.33	200	2.72

No.	Inbred line	DSS	Reaction type	Incidence (%)	Lesion length (cm)	Lesion width (cm)	TSW	Yield (t/ha)
33	139-5-j	2.03	I	44.12	19.33	1.07	237	1.8
34	30H83-561-1-1-1-1	2.22	I	50.98	13.00	0.83	203	2.93
35	SZSYNA-99-F2-80-3-4-1	2.81	S	76.47	27.67	1.83	237	3.38
36	124-b(109)	3.03	S	82.84	19.33	1.33	270	5.59
37	F7215	2.5	I	62.75	21.33	1.07	273	2.92
38	136-a	1.75	R	42.16	17.33	1.33	363	3.62
39	DE-78-Z-126-3-2-1-2-1	2.56	S	65.20	19.00	1.00	237	3.54
40	Gibe-1-186-2-2-1	2.81	S	49.02	19.33	1.33	360	2.3
41	Pool9A-128-5-1-1-1	2.67	S	68.36	15.33	1.67	223	2.87
42	30H83-71-5-1-1-1-1	1.94	R	50.00	16.67	2.00	193	2.84
43	SZSYNA-99-F2-3-6-2-1	2.28	I	63.24	15.33	2.00	233	2.36
44	SC-715-13-2-1	2.47	I	66.67	16.33	1.17	210	2.34
45	SC-22-430(63)	3.08	S	89.71	14.33	1.67	227	2.05
46	Kuleni-C1-101-1-1-1	2.81	S	68.63	7.67	1.17	237	3.07
47	Iloo'E-1-12-4-1-1	2.17	I	33.33	10.67	1.33	243	1.57
48	(DRB-F2-60-1-2)-B-1-B-B-B	2.72	S	67.65	21.00	2.00	230	2.48
49	142-1-e	1.81	R	49.51	16.33	1.50	287	4.29
50	CML 197	3.39	S	98.53	19.67	2.67	213	4.41
	LSD	0.396	—	19.159	9.013	0.902	47	1.3
	Pr > f	**	—	**	**	*	**	**
	CV (%)	10.1	—	18.8	32.1	36.9	11.9	24.7
	Overall mean	2.43	—	62.93	17.31	1.51	245	3.23

DSS = disease severity score (0.00–5.00); R = resistant (1.0–2.00); I = intermediate (2.10–2.50); susceptible (2.51–5.00); and TSW = thousand seed weight.

*; ** = Significant at 0.05 and 0.01 probability levels, respectively.

Table 4. Maize leaf blight reactions, grain yield, and thousand seed weight of 50 inbred lines tested during 2012 at Bako research Center in Ethiopia.

observed for CML 197, Kuleni-C1-101-1-1-1, and SC-22-430(63), suggesting that they were susceptible to the disease. The final severity score and AUDPC values provided sufficient estimation of the reaction of the inbred lines to *E. turcicum*. The inbred lines that were classified as resistant showed significantly lower AUDPC values than the susceptible ones (**Figure 1**). Furthermore, susceptible inbred lines tended to show a rapid increase in severity of the disease compared with the resistant lines culminating in higher severity scores toward maturity unlike the resistant ones. The severity of the disease was slightly higher in 2011 than 2012 (**Tables 3** and **4**). This was likely due to the low rainfall that was received at flowering in 2012,

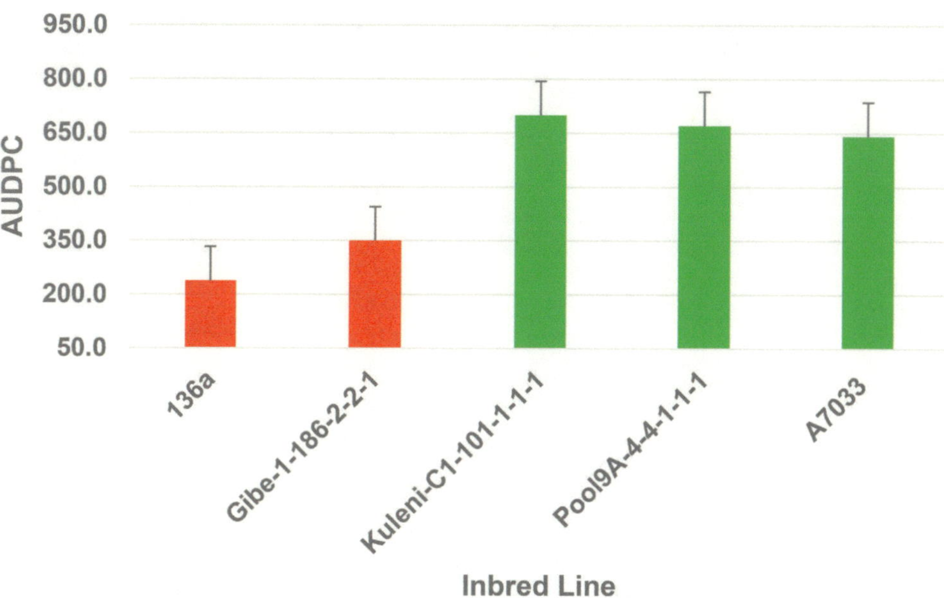

Figure 1. Area under disease progress curve for resistant (red) and susceptible (green) maize inbred lines inoculated with isolates of *E. turcicum* in the field.

which was not conducive for the development of the disease. Nonetheless, the environmental conditions were generally favorable for leaf blight development during the two testing seasons. Previous studies involving leaf blight showed that the dropper inoculation was efficient and minimized the chances of disease escape from evaluation [9]. In this study, the inoculation technique was easy to employ and reliable. There were clear differences between resistant and susceptible genotypes, and at the flowering stage, the later genotypes exhibited a moderate increase in diseased leaf tissue. In some cases, relatively less susceptible individual genotypes were identifiable. The selection of such less susceptible genotypes can result in the accumulations of minor genes that can elevate the level of field resistance [35–37].

3.2. Genetic polymorphism

The twenty SSR primers identified 108 alleles among the 20 maize inbred lines. Between 1 to 11 alleles were scored across the SSR loci (**Table 5**). Two loci (Phi 037, Umc1296) each revealed only a single allele. The maximum number of alleles (11) was detected at the Bnlg 2190 locus. The maximum PIC estimated for all loci was 0.8028 with a mean of 0.54 (**Table 5**). The expected heterozygosity (He) values, as a measure of allelic diversity at a locus, varied from 0.0000 to 0.8395 with an average of 0.5774. These values were well correlated with the number of alleles. Ten SSR loci (Umc1568, Nc003, Umc2214, Umc2038, Phi085, Umc1153, Bnlg238, Phi054, Bnlg2190, and Bnlg240) attained a PIC value >0.6, which indicated their potential to detect differences between the inbred lines.

The genetic diversity of the germplasm is one of the most important factors limiting the number of alleles identified per microsatellite locus during screening. However, other factors

SSR locus	Repeat types	Bin number	Number alleles	PIC value	He
Umc1568	TCG	1.02	6	0.6833	0.7250
Bnlg176	___	1.03	4	0.3092	0.3378
Bnlg182	___	1.03	6	0.5510	0.5888
Phi 037	AG	1.08	1	0.0000	0.0000
Bnlg 108	___	2.04	4	0.4253	0.4637
Nc003	AG	2.06	6	0.7429	0.7778
Umc2214	CTT	2.1	8	0.7075	0.7350
Bnlg602	___	3.04	6	0.4701	0.4900
Umc2038	GAC	4.06	4	0.6311	0.6925
Phi085	AACGC	5.06	4	0.6695	0.7222
Umc1153	TCA	5.09	8	0.6683	0.7036
Bnlg238	___	6	8	0.7689	0.7922
Umc1296	GGT	6.07	1	0.0000	0.0000
PhiI015	AAAC	8.08	7	0.5112	0.5938
Umc1367	CTG	9.05	2	0.4949	0.5850
Phi054	AG	10.03	6	0.8028	0.8255
Umc1677	GGC	10.05	7	0.3047	0.3750
Bnlg2190	AG	10.06	11	0.8224	0.8395
Bnlg240		8.06	7	0.7777	0.8025
umc2361	CCT	8.06	2	0.3743	0.4986

PIC = polymorphic information content and He = heterozygosity.

Table 5. Information about the 20 SSR loci used in this study.

such as the number of SSR loci and repeat types as well as the methodologies employed for the detection of polymorphic markers have been reported to influence allelic differences. In this work, the mean number of alleles (5.4) was in agreement with those reported in maize [38]. Similarly, values of number of SSR loci used in this study closely agreed with the findings reported previously [13, 39]. In addition, the mean PIC value determined in the present investigation was in agreement with the findings that were obtained in earlier studies that involving the use of SSR markers on maize inbred lines [40, 41]. The PIC value demonstrates the usefulness of the SSR loci and their potential to detect differences among the inbred lines based on their genetic relationships. The dinucleotide SSR loci (phi054, nc003, bnlg2190) identified the largest mean number of alleles (7.67) and mean PIC (0.79), as compared to tri-, tetra-, and penta-nucleotide repeats in the study, which was in close agreement with previous observations in maize [40, 42].

In this study, automated analysis was used for screening the microsatellites, resolving allelic variation better than using gel electrophoretic analysis for instance. This may be particularly important for SSR loci containing dinucleotide repeats whose amplification products are between 130 and 200 bp, because PCR products differing by two base pairs cannot be resolved with agarose gel electrophoresis [40, 43].

The ability to measure genetic distances between the inbred lines that reflect pedigree relationship ensures a more stringent evaluation of the adequacy of marker profile data; hence, the minimum genetic distance which was revealed between CML-202 and I100E-1-9-1-1-1-1-1 (0.28) was a good indication, confirming the power of SSR markers to distinguish closely related inbred lines. Similar findings were reported for maize inbred lines using SSR markers [44–46].

3.3. Cluster analysis

The dendrogram obtained using the UPGMA clustering algorithm based on SSR data matrices grouped the inbred lines into five categories (**Figure 2**). This information, in combination with the pedigree records and combining ability tests, will be valuable for selecting (or identifying) optimal crosses and assigning inbred lines into heterotic groups. The greatest distance was found between the cluster containing the inbred line CML-202 line and the cluster of the inbred line Gibe-1-91-1-1-1-1. Cluster I consisted of inbred lines that are adapted to mid-altitude as well as some originating from CIMMYT. Most of the mid-altitude inbred lines in this group originated from the heterotic group Kitale Synthetic II and constitute the largest group in the cluster. In Cluster II, CIMMYT inbred lines CML312

Figure 2. Dendrogram showing genetic relationship among 20 maize inbred lines tested using 20 SSR markers. The five clusters among the inbred lines are denoted from I to V.

and CML395 were grouped along with two local inbred lines, with two subdivisions in the main group. Cluster III contained two major subgroups, one containing CIMMYT inbred lines and the other containing local inbred lines. In terms of pedigree, these inbred lines are closely related and belong to the heterotic group AB, thus supporting the observation of a positive relationship between the pedigree and the SSR marker groupings in this study. In another cluster, two CIMMYT inbred lines (CML-443 and CML-197) were grouped closely, as revealed on the UPGMA dendrogram (**Figure 2**). These two inbred lines were also grouped in the same heterotic groups A and AB, based on their heterosis. Cluster V consisted of one CIMMYT inbred line and two locally adapted mid-altitude inbred lines. The separation of these elite mid-altitude maize inbred lines into genetically distinct groups may be associated with high heterotic response and increased combining ability useful for hybrid development.

The majority of the inbred lines (60%) that were evaluated in this study were previously developed by the national maize breeding program in Ethiopia. Because of the potential of encountering genetic admixtures or incomplete pedigree records in breeding programs, discrepancies in classification of germplasm may occur when comparing molecular results with classification based on pedigree relatedness. The effects of selection, genetic drift, and mutation may contribute to these discrepancies. The technique of clustering inbred lines can create apparent discrepancies, when one inbred line that is related to two inbred lines from separate clusters is then grouped with the inbred to which it is more closely related [40, 47]. Nonetheless, the SSR markers separated most of the inbred lines into distinguishable clusters, which generally agreed with the existing pedigree records and the findings that were reported previously [27, 42].

4. Conclusions

The inbred lines showed significant differences in reaction to the leaf blight disease and were classified into three categories namely resistant, intermediate, or susceptible. The mean disease severity and upper leaf area infection varied from 2.04 to 3.25 and 3.3% to 100% respectively. Seven inbred lines were identified as potential sources of resistance to leaf blight for the genetic improvement of maize under the mid-altitude agroecology in Ethiopia. The genotyping detected 108 alleles and grouped the inbred lines into five clusters consistent with their pedigrees. The genetic grouping present in the population as determined in this study will be useful in the exploitation of tropical germplasm for hybrid maize breeding programs. The inbred lines that were identified as resistant to leaf blight can be considered as source material for disease resistance under the mid-altitude agroecological conditions in Ethiopia. The genetic grouping of the inbred lines was valuable information for future maize breeding programs. The use of SSR markers was able to provide complimentary information regarding the relatedness of the elite inbred lines that were evaluated. The high PIC value across all loci was strong evidence confirming the potential for SSR markers to discriminate between inbred lines of diverse sources and even between closely related genotypes. A number of loci that were identified with high PIC values indicated their usefulness for diversity analysis of maize inbred lines. The approach used in the study enables clear differentiation between inbred lines and their classification into distinct groups based on genetic distance estimates generated through selected polymorphic SSR primers.

There will be merit in establishing resistance breeding program aimed at developing varieties with increased adult plant resistance to TLB in Ethiopia. Such varieties offer one of the most effective and affordable ways to overcome the problem of leaf diseases of maize in the mid-altitude agroecology in Ethiopia and similar environments in SSA. Therefore, further testing of the resistant germplasm identified in this study across more locations and seasons will also be merited.

Author details

Abera Wende[1], Hussein Shimelis[1] and Eastonce T. Gwata[2]*

*Address all correspondence to: ectgwata@gmail.com

1 African Centre for Crop Improvement, University of KwaZulu-Natal, Scottsville, South Africa

2 Department of Plant Production, University of Venda, Thohoyandou, Limpopo, South Africa

References

[1] FAO. World Agricultural Production [Internet]. 2011. Available from: http://faostat.fao.org/default.aspx [Accessed: 14 June 2013]

[2] FAO. World Agricultural Production [Internet]. 2007. Available from: http://faostat.fao.org/default.aspx [Accessed: 30 November 2011]

[3] Rosegrant MW, Paisner MS, Meijer S, Witcover J. Strategies for selection of host resistance. The African Crop Science Journal. 2001;**1**:189-198

[4] Mosisa W, Legesse W, Berhanu T, Girma D, Girum A, Wende A, Tolera K, Gezahegn B, Dagne W, Solomon A, Habtamu Z, Kasa Y, Temesgen C, Habte J, Demoz N, Getachew B. Status and future direction of maize research and production in Ethiopia. In: Proceedings of the 3rd National Maize Workshop of Ethiopia; 18-20 April 2011; Addis Ababa, Ethiopia. 2012. pp. 17-23

[5] Girma D, Solomon A, Emana G, Ferdu A. Review of the past decade's (2001-2011) research on pre-harvest insect pests of maize in Ethiopia. In: Proceedings of the 3rd National Maize Workshop of Ethiopia; 18-20 April 2011; Addis Ababa, Ethiopia. 2012. pp. 166-173

[6] Tewabech T, Dagne W, Girma D, Meseret N, Solomon A, Habte J. Maize pathology research in Ethiopia in the 2000s: A review. In: Proceedings of the 3rd National Maize Workshop of Ethiopia; 18-20 April 2011; Addis Ababa, Ethiopia. 2012. pp. 193-201

[7] Juliana BO, Marco AG, Isaias OG, Luis EAC. New resistance genes in *Zea mays Exserohilum turcicum* pathosystem. Genetics and Molecular Biology. 2005;**28**:435-439

[8] Cardwell KF, Schulthess F, Ndehma R, Ngoko Z. A systems approach to assess crop health and maize yield loss due to pests and diseases in Cameroon. Agriculture, Ecosystems & Environment. 1997;**65**:33-47

[9] Muiru WM, Mutitu WE, Kemenju JW. Reaction of some Kenyan maize genotypes to Turcicum leaf blight under green house and field conditions. Asian Journal of Plant Sciences. 2007;**6**:1190-1196

[10] Dagne W, Habtamu Z, Demissew A, Temam H, Harjit S. Combining ability of maize inbred lines for grain yield and reaction to grey leaf spot disease. The East African Journal of Sciences. 2008;**2**:135-145

[11] Njuguna JAM, Kendera JG, Muriithi LMM, Songa S, Othiambo RB. Overview of maize diseases in Kenya. Maize Review Workshop in Kenya. Kakamega, Kenya; 1990. pp. 45-51

[12] Welz HG, Geiger HH. Genes for resistance to northern corn leaf blight in diverse maize populations. Plant Breeding. 2000;**119**:1-14

[13] Legesse WB, Myburg AA, Pixley KV, Botha AM. Genetic diversity of maize inbred lines revealed by SSR markers. Hereditas. 2007;**144**:10-17

[14] Singh R, Mam VP, Koranga KS, Bisht GS, Khandelwal RS, Bhandar P, Pan SK. Identification of additional sources of resistance to *Exserohilum turcicum* in maize. SABRAO Journal of Breeding and Genetics. 2004;**36**:45-47

[15] Ogiliari JB, Guimaraes MA, IO GI. New resistance genes 400 in the *Zea mays – Exserohilum turcicum* pathosystem. Genetics and Molecular Biology. 2005;**28**:435-439

[16] Yeshitela D. Cloning and Characterization of Xylanase Genes from Phytophathogenic Fungi with a Special Reference to *Helminthosporium turcicum* the Cause of Northern Leaf Blight of Maize [Thesis]. Finland: University of Helsinki; 2003

[17] Gevers HO. A new major gene for resistance to *Helminthosporium turcicum* leaf blight of maize. Plant Disease. 1997;**59**:296-299

[18] Leonard KJ. Proposed nomenclature for pathogenic races of *Exserohilum turcicum* on corn. Plant Disease. 1989;**73**:776-777

[19] Adipala E. Reaction of 13 maize genotypes to *E. turcicum* in different agro-ecological zones of Uganda. East African Agricultural and Forestry Journal. 1994;**59**:213-218

[20] Pratt RC, Gordon K, Lipps P, Asea G, Bigrawa G, Pixley K. Use of IPM in the control of multiple diseases of maize. The African Crop Science Journal. 2003;**11**:189-198

[21] Terron A, Preciado E, Cordova H, Mickelson H, Lopez R. Determinacion del patron heterotico de 30 lineas de maiz derivadas del la poblacion 43 SR del CIMMYT. Mesoamerican Agronomy. 1997;**8**:26-34

[22] Melchinger AE, Gumber RK. Overview of heterosis and heterotic groups in agronomic crops. In: Lamkey KR, Staub JE, editors. Concepts and Breeding of Heterosis in Crop Plants. Madison, USA: CSSA; 1998

[23] Reif JC, Melchinger AE, Xia XC, Warburton ML, Hoisington DA, Vasal SK, Beck D, Bohn M, Frisch M. Use of SSRs for establishing heterotic groups in subtropical maize. Theoretical and Applied Genetics. 2003;**107**:947-957

[24] Lee M, Godshalk EB, Lamkey KR, Woodman WW. Association of restriction fragment length polymorphisms among maize inbreds with agronomic performance of their crosses. Crop Science. 1989;**29**:1067-1071

[25] Livini C, Anjmone-Marsan P, Melchinger AE, Messmer MM, Motto M. Genetic diversity of maize inbred lines within and among heterotic groups revealed by RFLPs. Theoretical and Applied Genetics. 1992;**84**:17-25

[26] Dubreuil P, Dufour P, Krejci P, Causse M, Vienne D, Gallais A, Charcosset S. Organization of RFLP diversity among inbred lines of maize representing the most significant heterotic groups. Crop Science. 1996;**36**:790-799

[27] Pejic I, Ajmone-Marsan P, Morgante M, Kozumplick P, Castiglioni P, Taramino G, Motto M. Comparative analysis of genetic similarity among maize inbred lines detected by RFLPs, RAPDs, SSRs, and AFLPs. Theoretical and Applied Genetics. 1998;**97**:1248-1255

[28] Wende A, Shimelis H, Derera J, Worku M, Laing M. Heterosis and combining ability of elite maize inbred lines under northern corn leaf blight disease prone environments of the mid-altitude tropics. Euphytica. 2016;**208**:391-400

[29] Campbell CL, Madden LV. Introduction of Plant Disease Epidemiology. New York, USA: Wiley; 1991

[30] Payne RW, Murray DA, Harding SA, Baird DB, Soutar DM. GenStat for Windows (11th Edition) Introduction. VSN International: Hemel Hempstead; 2008

[31] Mbogori MN, Kimani M, Kuria A, Lagat M, Danson JW. Optimization of FTA technology for large scale plant DNA isolation for use in marker-assisted selection. The African Journal of Biotechnology. 2006;**5**:693-696

[32] Smith JSC, Chin ECL, Shu H, Smith OS, Wall SJ, Senior ML, Mitchell SE, Kresovitch S, Ziegle J. An evaluation of the utility of SSR loci as molecular markers in maize: Comparison with data from RFLP and pedigree. Theoretical and Applied Genetics. 1997;**95**:163-173

[33] Muriithi LMM. Response of maize genotypes to *Turcicum* blight and other economically important diseases. In: Proceedings of the 3rd KARI Scientific Conference; 7-9 October 1992; Nairobi, Kenya

[34] Levy Y. Variation in fitness among field isolates of *Exserohilum turcicum* in Israel. Plant Disease. 1991;**75**:163-166

[35] Bowen KL, Pedersen WL. Effects of northern leaf blight components on corn inbreds. Plant Disease. 1988;**72**:952-956

[36] Ceballos H, Deutsch JA, Gutiérrez H. Recurrent selection for resistance to *Exserohilum turcicum* in eight subtropical maize populations. Crop Science. 1991;**31**:964-971

[37] Ojulong HF, Adipala E, Rubaihayo PR. Diallel analysis for reaction to *Exserohilum turcicum* of maize cultivars and crosses. The African Crop Science Journal. 1996;**4**:19-27

[38] Vaz Patto MC, Satovic Z, Pêgo S, Fevereiro P. Assessing the genetic diversity of Portuguese maize germplasm using microsatellite markers. Euphytica. 2004;**137**:63-72

[39] Pinto LR, Vieira MLC, de Souza CL, de Souza AP. Genetic diversity assessed by microsatellites in tropical maize population submitted to high-density reciprocal recurrent selection. Euphytica 2003;**134**:277-286

[40] Senior ML, Murphy JP, Goodman MM, Stuber CW. Utility of SSRs for determining genetic similarities and relationships in maize using an agarose gel system. Crop Science. 1998;**38**:1088-1098

[41] Heckenberger M, Bohn M, Zeigle JS, Joe LK, Hauser JD, Hutten M, Melchinger AE. Variation of DNA fingerprints among accessions within maize inbred lines and implications for identification of essentially derived varieties. Genetic and technical sources of variation in SSR data. Molecular Breeding. 2002;**10**:181-191

[42] Enoki H, Sato H, Koinuma K. SSR analysis of genetic diversity among maize inbred lines adapted to cold regions of Japan. Theoretical and Applied Genetics. 2002;**104**:1270-1278

[43] Sibov ST, de Souz CL, Garica AF, Silva AR, Mongolin CA, Benchimol LL, de Souza AP. Molecular mapping in tropical maize (*Zea mays L.*) using microsatellite markers. Map construction and localization of loci showing distorted segregation. Hereditas. 2003;**139**:96-106

[44] Boppenmeier J, Melchinger AE, Brunklaus-Jung E, Geiger HH, Herrmann RG. Genetic diversity for RFLP in European maize inbreds: In relation to performance of Flint x dent crosses for forage traits. Crop Science. 1992;**32**:895-902

[45] Lanza LLB, de Souza Jr CL, Ottoboni LM, Vieira MLC, de Souza AP. Genetic distance of inbred lines and prediction of maize single cross performance using RAPD markers. Theoretical and Applied Genetics. 1997;**94**:1023-1030

[46] Li Y, Du J, Wang T, Shi Y, Song Y, Jia J. Genetic diversity and relationships among Chinese maize inbred lines revealed by SSR markers. Maydica. 2002;**47**:93-101

[47] Ajmone-Marsan P, Castiglioni P, Fusari F, Kuiper M, Motto M. Genetic diversity and its relationship to hybrid performance in maize as revealed by RFLP and AFLP markers. Theoretical and Applied Genetics. 1998;**96**:219-227

Chapter 5

Use of Technology to Increase the Productivity of Corn in Brazil

Wilian Henrique Diniz Buso and
Luciana Borges e Silva

Additional information is available at the end of the chapter

http://dx.doi.org/10.5772/intechopen.70808

Abstract

Brazil is one of the world's principal producers of corn, and over the past few decades, a range of new technologies have been incorporated to guarantee advances in the productivity of this crop. Initially, genetic enhancement was achieved through the production of hybrid seed that was more productive than freely pollinated cultivars. Subsequent adjustments to cultivation practices, such as the reduction in row spacing, balanced fertilisation and direct planting, have contributed to a progressive increase in productivity. The authorisation of the marketing of transgenic seed, providing resistance to insect pests and herbicides, contributed further to productivity by reducing losses to pests (*Spodoptera frugiperda*) and competition with weeds. Together, all these technological advances have contributed to ever increasing gains in the productivity of Brazilian corn crops.

Keywords: biotechnology, management, research, *Zea mays*

1. Introduction

Brazil covers a total area of 8,511,996 km^2, divided into five geographic regions characterised by major climatic and economic differences [1]. The equatorial northern region has a rainy climate, and is covered by the world's largest area of pristine tropical rainforest, while the Northeast is mostly semi-arid with some irrigated areas. The Midwest, Southeast and South are the principal grain-producing regions.

Corn (*Zea mays*) is the grain cultivated in the largest volume worldwide, with the United States, China and Brazil being the principal producers. In Brazil, 15,922.5 million hectares were planted with corn in the 2015–2016 season, with a mean productivity of 4178 kg ha^{-1},

rising to 16,772 million hectares in 2016–2017, with an expected mean productivity of 5305 kg ha^{-1}, with a total harvest of 88,969.40 million tons [2].

The technological advances in the production of corn in Brazil involved the correction of the soil (acidity, neutralisation of aluminium and increase in base saturation). Over the subsequent years, direct planting was adopted as a strategy for the protection of the soil, using corn stover as a way of increasing the amount of organic matter in the soil. Subsequent research tested the reduction of the spacing of the rows from 0.90 to 0.45–0.50 m to optimise the performance of seeding machines and improve the density of plantations, leading to an improvement in the absorption of soil nutrients by the roots of the plants.

The reduction in spacing also contributed to an improvement in the control of weeds, through the more rapid formation of ground cover and shading of the soil, in addition to an increase in the efficiency of fertilisers. Subsequently, the introduction of genetically modified organisms for the control of the fall armyworm (*Spodoptera frugiperda*), the principal insect pest of corn plantations, also resulted in gains in productivity.

Recent advances in biotechnology have included the incorporation of a number of proteins derived from the bacterium *Bacillus thuringiensis* to control of a range of insect pests (*Elasmopalpus lignosellus, Agrotis ipsilon, S. frugiperda, Diatraeia saccharalis* and *Helicoverpa zea*), reducing damage to the plants, and improving productivity. The subsequent introduction of hybrids resistant to insects and herbicides (glyphosate and ammonium glufosinate) has further reduced the costs of the control of insect pests and weeds. The combination of these technologies has brought significant gains in the productivity of corn, both in Brazil, and the rest of the world.

2. Technologies adopted to increase productivity

2.1. Brazilian research in corn production

In Brazil, the corn seed industry involves a number of national and multinational corporations, as well as public entities that are all working to develop new cultivars and technologies [3]. In recent years, these enterprises have marketed cultivars that target specific productive sectors, which rely on high, medium and low levels of technology. The former two sectors use hybrids, while the low technology sector still relies on many freely pollinated varieties.

Araújo et al. [4] investigated the collaborative public networks of corn research in Brazil between 2006 and 2010, and found close ties between the institutions involved in the enhancement of cultivars and those working on the development of technology for the improvement of productivity (**Figure 1**). Research efforts are concentrated in Southeast Brazil, where the Brazilian Public Agricultural Research Corporation (EMBRAPA) and São Paulo State University (UNESP) have close links with a number of other research institutions, developing collaborative research projects for the divulgation of new technologies for corn production.

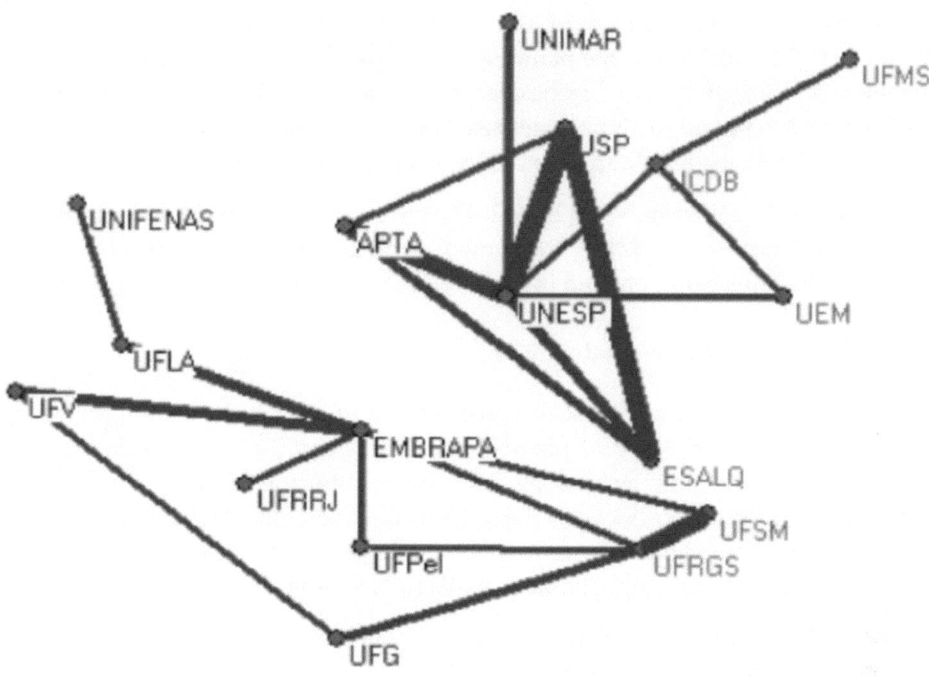

Figure 1. Nucleus of the collaborative public corn research network in Brazil (2006–2010). Source: Araújo et al. [4].

Galvão et al. [5] evaluated the advances in the production of corn in Brazil since the 1940s and found that technology has contributed to an increase in productivity of 379% over the past 70 years. Research institutions have contributed to this increase in productivity through the development of research, cultivars and technologies, the training of specialised personnel, and the communication of information to farmers. This technological development has resulted in Brazil reaching third place in the world ranking of corn producers and exporters, with total production increasing from 5.6 million tons in 1944 to more than 89 million tons in 2017.

2.2. Use of biotechnology

The Brazilian National Technical Commission for Biosecurity (NTCBIO) was created by federal decree number 1520/95. This organ is responsible for the development of legal norms on the biosecurity of genetically modified organisms (GMOs) and the classification of their potential risks. The commission was initially responsible for the authorisation of experiments on transgenic plants in Brazil. The cultivation of genetically modified plants in Brazil began in the 1990s with the illegal introduction of the Roundup Read (RR) soybean, which is resistant to the herbicide glyphosate, in the state of Rio Grande do Sul.

In the specific case of transgenic corn, the importation of seed from Argentina was first authorised in 2005, in an attempt to overcome the poor harvest of this year. Eventually, in May 2007, the NTCBIO authorised the sale of transgenic corn in Brazil. Currently, most areas planted with corn in Brazil involve some transgenic variety, and the vast majority of hybrids are now resistant to insects (lepidopterans) and herbicides (glyphosate and ammonium glufosinate). In 2007, the NTCBIO authorised the planting of *Bt* corn, which contains the protein cry1fAb for the control of *S. frugiperda* and *Diatraeia saccharalles*, and in 2008, it permitted the sale of RR corn seed, which is resistant to glyphosate-based herbicides, as an alternative for the management of weeds, due to the ample spectrum of control of these plants.

In the most recent Brazilian harvest (2016–2017), transgenic corn, resistant to insects and/or herbicides, should account for 82% of the summer crop and 92% of the second planting, with transgenic hybrids thus being planted in more than 88% of the total area cultivated.

2.3. Use of hybrids to increase productivity

Tollenaar and Lee [6] concluded that the productivity of corn is dependent on the specific genetic characteristics of the hybrid planted, favourable environmental conditions and the adoption of adequate farming technology. The potential for the production of grain will be influenced by the interaction between the hybrid and the cultivation conditions, with the same hybrid responding differently to distinct conditions, depending on the ambient temperatures, the incidence of sunlight and the availability of water.

Each year, a number of new hybrids are marketed, following extensive testing in the principal corn-producing regions of the country to determine the conditions to which the hybrid is best adapted. In a study of 22 hybrids at 14 different sites, Cardoso et al. [7] observed varying responses, with some cultivars being well-adapted to a wide range of conditions, in which they maintain their productivity, whereas others are better adapted to certain specific conditions.

In a study of 10 hybrids during 3 different planting periods (18/11/2011, 31/01/2012 and 20/02/2012), Buso and Arnhold [8] recorded variation in the performance of the cultivars under different seasonal conditions. In this analysis (**Figure 2**), the hybrid AGN 30A77H performed better than all the other hybrids in the first two periods (18/11/2011 and 31/01/2012), whereas the third period (20/02/2012) was found to be unfavourable due to water stress.

Sousa et al. [9] evaluated 36 corn hybrids cultivated under water stress and found that the performance of these cultivars varied according to the humidity of the soil, with some hybrids performing much better than others under these extreme conditions. The testing of these hybrids contributed to the identification of the cultivars best adapted to the second planting in Brazil, principally under conditions of water stress, in the different Brazilian regions. The interim harvest is planted between January and March. Silva et al. [10] noted that, due to the precocity of the hybrids, the interim crop is favoured by the fact that the flowering period coincides with the rainy season, when more groundwater is available, contributing to productivity.

2.4. Changes in production management

In addition to genetic enhancement and the use of biotechnology, other agricultural practices contributed to the increase in corn productivity, such as nutrient management, the reduction

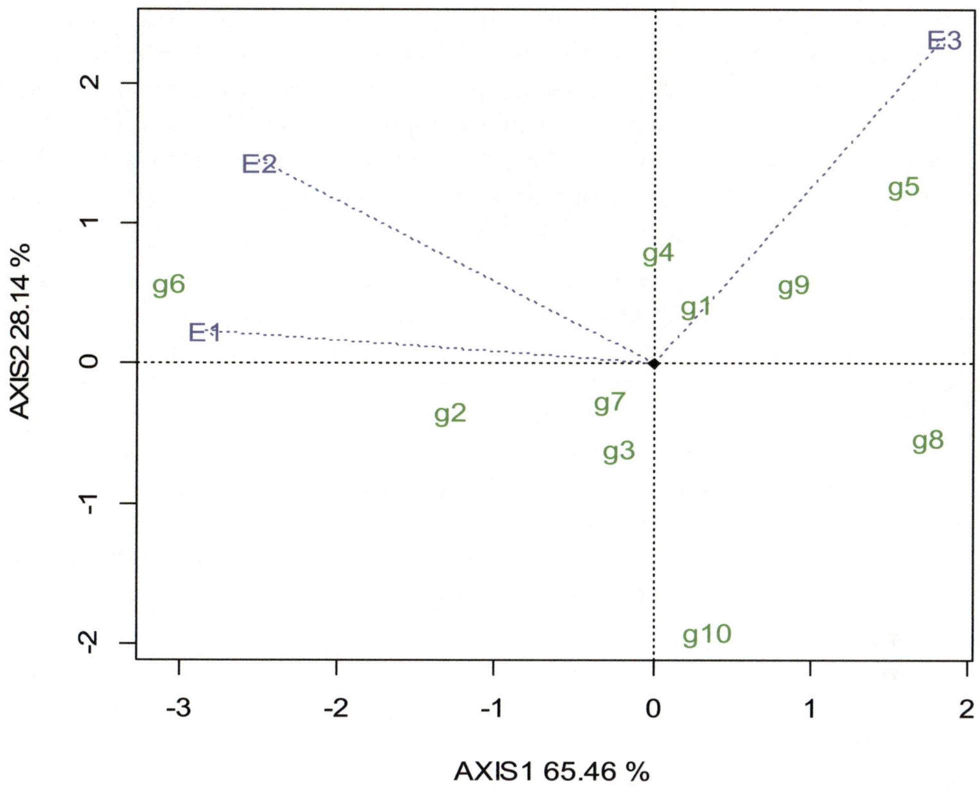

Figure 2. Graph produced in GGEbiplot showing the perspective of different hybrids in three distinct seeding periods, E1 (18/11/2011), E2 (31/01/2012) and E3 (20/02/2012). Codes: g1 = Truck, g2 = Formula, g3 = P30F53, g4 = P3646H, g5 = P30F35H, g6 = AGN 30A77H, g7 = AGN 30A37H, g8 = AG 8088 PRO, g9 = DKB 390 and g10 = DKB Bi9440. Source: Buso and Arnhold [8].

in row spacing, adjustments of plant density and the use of direct seeding. The adjustment of the spatial arrangement of the plants (in particular the density of the plantation) and the reduction of row spacing had positive effects on productivity, through the increase in the incidence of sunlight and the better exploitation of the environment by the genotype [11]. The increase in population density results in gains in productivity up to an optimum number of plants per unit area, which varies according to the hybrid and the environmental conditions, with productivity decreasing at densities above this optimum level [12]. Increasing the density of plants leads to an increase in the competition among plants for water, nutrients, sunlight and CO_2 [13], and may also induce sterility and reduce the amount of grain per cob, resulting in a loss of productivity.

In their analysis of different row spacing parameters and population densities (**Table 1**), Farinelli et al. [14] observed that productivity was influenced by the reduction in spacing and the increase in the density of seeding, with the highest productivity (7842 kg ha^{-1}) being obtained with the most reduced spacing, of 40 cm (**Table 1**). This result may be related to the increased efficiency of the plants in the interception of sunlight, and a decrease in the competition for sunlight, water and nutrients among the plants in the same row. These authors also recorded an increase in productivity with increasing population density, up to 80,000 plants ha^{-1} (**Table 1**). These gains in productivity accruing to increasing population density are related to the use of hybrids better adapted to high population densities. These hybrids are smaller, have more erect leaf architecture, rapid emission of the style-stigma, coordination of the anthesis with the emission of the stigmas, rapid development of the first cob, reduced size of the tassel and an even greater efficiency in the production of grain per unit area.

Silva et al. [15] found that a row spacing of 0.45 m resulted in a 17% gain in productivity in comparison with a 0.90 m spacing (**Table 2**), and found many other studies with similar results, showing that considerable gains can be obtained by reducing the 0.90 m row spacing that had been used for many years. These authors also found that densities of 60,000 and 80,000 plants ha^{-1} resulted in gains in productivity of 12.5 and 13.6%, respectively, in comparison with the more traditional density of 40,000 plants ha^{-1} (**Table 2**). These results indicate that the hybrids tested tolerate an increase in planting density without affecting productivity. However, the density of 60,000 plants ha^{-1} appears to be the most viable option, considering that the gain in productivity is only negligibly lower from that at 80,000 plants ha^{-1}, while the adoption of a greater plant density implies higher costs for the purchase of seed.

In an analysis of the harvests of 2 years, Buso et al. [16] recorded different patterns of productivity between years for different parameters of row spacing and planting density (**Table 3**). In the first year, productivity was greater at the higher densities (70,000 and 80,0000 plants ha^{-1}), with 10,922–11,796 kg ha^{-1}, while the lower density (60,000 plants ha^{-1}) produced only 9118 kg ha^{-1}. In the second year, the middle density (70,000 plants ha^{-1}) was significantly more

Spacing (m)	**Productivity (kg ha^{-1})**	**Density (plants ha^{-1})**	**Productivity (kg ha^{-1})**
0.4	7842 a	40,000	6320 b
0.6	7372 ab	60,000	7777 a
0.8	6974 b	80,000	8091 a

The mean values in the same column followed by different letters are significantly different ($p \leq 5\%$) from each other, based on Tukey's test. Adapted from Farinelli et al. [14].

Table 1. Productivity of corn according to different row spacing and plant densities.

Spacing (m)	**Productivity (kg ha^{-1})**	**Density (plants ha^{-1})**	**Productivity (kg ha^{-1})**
0.45	8514 a	40,000	7256 b
0.90	7263 b	60,000	8163 a
–	–	80,000	8246 a

The mean values in the same column followed by different letters are significantly different ($p \leq 5\%$) from each other, based on Tukey's test. Adapted from Silva et al. [15].

Table 2. Productivity of corn under different standards of row spacing and plant density.

productive (6253 kg ha^{-1}) than either of the other densities, with 60,000 plants ha^{-1} producing only 5045 kg ha^{-1} of corn and 80,000 plants ha^{-1} producing 5606 kg ha^{-1} (**Table 3**).

The reduction in row spacing contributes to gains in productivity through the optimal distribution of the plants per unit area and provides the best management strategy for the control of weeds, due to the rapid growth of the plants, which closes over the gaps and increases the interception of sunlight, impeding the growth of weeds. It also increases the exploitation of the soil by the root system of the plants, and reduces planting costs, given that the same machinery used to seed other crops, such as soybean, bean and sorghum, can be used to plant the corn, due to the fact that these crops use the same row spacing.

The majority of the 16 million hectares used to produce corn in Brazil are cultivated by direct planting [2]. However, the adequate management of the soil is essential to guarantee the efficiency of this system [17]. This requires mechanical-, edaphic- and vegetation-based conservation practices, in particular, the use of cover crops to form a layer of stover, increase the organic material and contribute to the greater retention of nutrients during the organic phase.

The maintenance of the surface stover is determined by the Carbon:Nitrogen (C:N) ratio and the lignin concentrations found in the different plant species used as cover and for the formation of the stover. Climatic conditions influence the velocity of the decomposition of the stover by microbial organisms, by determining the micro-environmental conditions for their development.

Carvalho et al. [17] studied the effects of cover crops and the successive cultivation of corn, and found that productivity was influenced by the type of stover, varying from 11,666 kg ha^{-1} (following wheat) to 12,780 kg ha^{-1} (following ruzi grass) during the 2010/2011 harvest (**Table 4**). Productivity was significantly higher for ruzi grass, brown hemp, Brazilian jackbean and pearl millet, in comparison with velvet bean and wheat. Productivity was highest in the context of the more accelerated decomposition of the residues of some of these species, which is associated with the quantity of dry matter produced. The chemical composition of the cover crops with the lowest concentrations of lignin, such as ruzi grass and Brazilian jackbean, and the production of greater volumes of dry matter may have favoured not only the quantity of nutrients, but also the synchrony of the liberation of the plantation for the seeding of the corn.

In general, the ruzi grass contributes to nutrient cycling and the excellent quality of the stover produced, which results in an increase in the levels of organic matter, protecting the soil from the direct impacts of erosive agents, as well as facilitating the management of weed growth. This grass also has a very aggressive root system, capable of recuperating nutrients that the planted crops are unable to access due to their depth in the soil profile.

The use of cover crops is essential to guarantee the sustainability of many different types of crops in all regions of Brazil, in particular those of the Cerrado domain, where the soils tend to be intensely weathered. In this case, the mineralization of the organic matter formed by the cover crops provides nutrients for the corn plantations. The most important nutrient for this crop (corn) is nitrogen, and the need for supplementation with this nutrient will depend on a series of factors, such as the history of the area and the crop planted before the corn, the definition of which will help define the optimum dosage, sources and the forms of nitrogen to be applied.

Harvest	Plant population (thousands ha^{-1})			Row spacing (m)	
	60	70	80	0.50	0.80
2010/2011	9118 aB	10,922 aA	11,796 aA	10,923 aA	10,301 aA
2011/2012	5045 bB	6253 bA	5606 bB	6437 bA	4831 bB

The mean values in the same row followed by different letters are significantly different ($p \leq 5\%$) from each other, based on the Scott-Knott test. Adapted from Buso et al. [16].

Table 3. Productivity of corn (kg ha^{-1}) in the 2010/2011 and 2011/2012 harvests for different plant densities and row spacing.

Cover crop	Level of N in the leaf (g kg^{-1})	Productivity (kg ha^{-1})
Ruzi grass (*Urochloa ruziziensis*)	26.0	12,780 a
Brown hemp (*Crotalaria juncea*)	27.1	12,710 a
Brazilian jackbean (*Canavalia brasiliensis*)	25.9	12,580 ab
Pigeon pea BRS mandarin (*Cajanus cajan*)	24.1	12,500 ab
Pearl millet 'BR05' (*Pennisetum glaucum*)	25.2	12,130 abc
Velvet bean (*Mucuna aterrima*)	26.4	11,750 c
Forage radish (*Raphanus sativus*)	28.8	12,280 abc
Sorghum 'BR 304' (*Sorghum bicolor*)	26.2	11,960 bc
Wheat (*Triticum aestivum*)	25.0	11,670 c
Native vegetation	24.4	11,940 c

The mean values in the same column followed by different letters are significantly different ($p \leq 5\%$) from each other, based on the Tukey-Kramer test. Adapted from Carvalho et al. [17].

Table 4. Level of N in the leaves of different cover crops and the productivity of the corn planted after these species.

Cover crop	Inoculated	Not inoculated
Crotalaria juncea	7795 b A	9124 a AB
Cajanus cajan	8299 b A	9338 a A
Pennisetum americanum	8487 a A	8159 a B
Pennisetum americanum + *Crotalaria juncea*	8632 a A	8569 a AB
Pennisetum americanum + *Cajanus cajan*	8164 a A	8796 a AB
Fallow	8288 a A	8153 a B

The mean values in the same row followed by different lower case letters, and in the same column by different upper case letters, are significantly different ($p \leq 5\%$) from each other, based on the Tukey-Kramer test. Adapted from Portugal et al. [18].

Table 5. Productivity (kg ha^{-1}) of corn from seed inoculated with the bacterium *Azospirillum brasilense* and seed not inoculated, raised following the planting of different cover crops.

One other management option, recommended by some authors, is the application of bacteria that contribute to the growth of the plants through a number of different mechanisms for the nitrogenous nutrition of the corn plantations. The most-studied crop-associated diazotophic bacteria are those of the genus *Azospirillum*. Portugal et al. [18] observed that inoculation of the corn seed with *Azospirillum* had different results, depending on the associated cover crop (**Table 5**). In this study, inoculation associated with *Crotalaria juncea* and *Cajanus cajan* did not result in any gains in productivity (**Table 5**), given that these two plants also fix nitrogen in the soil, benefiting the subsequent corn crop. In areas planted with grasses or left fallow, however, inoculation with *Azospirillum brasilense* tends to have an effect on productivity.

Author details

Wilian Henrique Diniz Buso* and Luciana Borges e Silva

*Address all correspondence to: wilian.buso@ifgoiano.edu.br

Instituto Federal de Ciência e Tecnologia Goiano Campus Ceres, Goiás, Brasil

References

[1] Oda LC, Soares BEC. Biotecnologia no Brasil. Aceitabilidade pública e desenvolvimento econômico. Parcerias Estratégicas. 2001;**10**:162-173

[2] Conab. Companhia Nacional de Abastecimento. Relatório de acompanhamento de safras. Available from: http://www.conab.gov.br/OlalaCMS/uploads/arquivos/17_03_14_15_28_33_boletim_graos_marco_2017bx.pdf [Accessed: March 31, 2017]

[3] Fuck MP, Bonacelli MB. A pesquisa pública e a indústria sementeira nos segmentos de sementes de soja e milho híbrido no Brasil. Revista Brasileira de Inovação. 2006;**6**:87-121

[4] Araújo UP, Prosdocimi MC, Gomes AF, Antonialli LM, Brito MJ. Características estruturais da rede colaborativa de pesquisa de milho no contexto das ciências agrárias. Revista Brasileira de Inovação. 2013;**12**:385-416

[5] Galvão JCC, Miranda GV, Trogello M, Fritsche-Neto R. Sete décadas de evolução do sistema produtivo da cultura do milho. Revista Ceres. 2014;**61**(Suppl):819-828

[6] Tollenaar M, Lee EA. Yield potential, yield stability and stress tolerance in maize. Field Crops Research. 2002;**75**:161-169

[7] Cardoso MJ, Carvalho HWL, Rocha LMP, Pacheco CAP, Guimarães LJM, Guimarães PEO, Parentony SN, Oliveira IR. Identificação de cultivares de milho com base na análise de estabilidade fenotípica no Meio-Norte brasileiro. Revista Ciência Agronômica. 2012;**43**:346-356

[8] Buso WHD, Arnhold E. Evaluation of corn hybrids under contrasting water availability conditions. Revista Caatinga. 2016;**29**:927-934

[9] Sousa RC, Bastos EA, Cardoso MJ, Ribeiro VQ, Brito RR. Desempenho produtivo de genótipos de milho sob déficit hídrico. Revista Brasileira de Milho e Sorgo. 2015;**14**:49-60

[10] Silva AG, Francischini R, Martins PDS. Desempenhos agronômico e econômico de cultivares de milho na safrinha. Revista Agrarian. 2015;**8**:1-11

[11] Demétrio CS, Fornasieri Filho D, Cazetta JO, Cazetta DA. Desempenho de híbridos de milho submetidos a diferentes espaçamentos e densidades populacionais. Pesquisa Agropecuária Brasileira. 2008;**43**:1691-1697

[12] Argenta GS, Silva PRF, Sangoi L. Arranjo de plantas de milho: análise do estadoda-arte. Ciência Rural. 2001;**31**:1075-1084

[13] Von Pinho RG, Gross MR, Steola AG, Mendes MC. Adubação nitrogenada, densidade e espaçamento de híbridos de milho em sistema plantio direto na região sudeste do Tocantins. Bragantia. 2008;**67**:733-739

[14] Farinelli R, Penariol FG, Fornasieri Filho D. Características agronômicas e produtividade de cultivares de milho em diferentes espaçamentos entre linhas e densidades populacionais. Científica. 2012;**40**:21-27

[15] Silva AF, Schoninger EL, Caione G, Kuffel C, Carvalho MAC. Produtividade de híbridos de milho em função do espaçamento e da população de plantas em sistema de plantio convencional. Revista Brasileira de Milho e Sorgo. 2014;**13**:162-173

[16] Buso WHD, Silva LB, Rios ADF, Firmiano RS. Corn agronomic characteristics according to crop year, spacing and plant population densities. Comunicata Scientiae. 2016;**7**:197-203. DOI: 10.14295/CS.v7i2.847

[17] Carvalho AM, Coser TR, Rein TA, Dantas RA, Silva RR, Souza KW. Manejo de plantas de cobertura na floração e na maturação fisiológica e seu efeito na produtividade do milho. Pesquisa Agropecuária Brasileira. 2015;**50**:551-561. DOI: 10.1590/S0100-204X2015000700005

[18] Portugal JR, Arf O, Peres AR, Gitti DC, NFS G. Coberturas vegetais, doses de nitrogênio e inoculação com *Azospirillum brasilense* em milho no Cerrado. Revista Ciência Agronômica. 2017;**48**:639-649

Chapter 6

Impacts of Nitrogen Fertilization and Conservation Tillage on the Agricultural Soils of the United States: A Review

Meimei Lin

Additional information is available at the end of the chapter

http://dx.doi.org/10.5772/intechopen.70550

Abstract

This review evaluated the effects of nitrogen (N) fertilization and conservation tillage systems on SOC stocks. N fertilizer additions had significant positive impact on SOC content, but the magnitude of this effect differed as a result of varying cropping systems: as cropping intensity increased, measured SOC content between fertilized and control treatment also increased. Significant differences of measured SOC stocks were detected between no till and conventional till, as well as reduced till and conventional till. However, no significant difference was observed between reduced till and no till. The differences of measured SOC content between no till and conventional till appeared to be significantly associated with treatment duration. Crop rotation system, soil texture, and mean annual precipitation did not have significant effects on SOC stocks produced from conventional tillage to no till. The results of this study confirmed that adoption of N fertilizer additions and conservational tillage systems can contribute to increased SOC level and thereby have the potential to mitigate the enhanced greenhouse gas effect. However, the evaluation of net carbon dioxide mitigation potential of these two recommended management practices should be carried out under a full carbon cycle analysis from carbon input to carbon output.

Keywords: soil organic carbon, nitrogen fertilization, no till, conventional till, agricultural productivity

1. Introduction

Modern agricultural practices, both agricultural extensification and intensification, have widespread negative environmental impacts such as biodiversity loss, damage to the environment, and degradation of critical ecosystem services [1]. Global climate change has been considered as one of the most pressing challenges that humans face in the 21st century [2]. As one of the major greenhouse gases (GHG), atmospheric CO_2 contributes substantially to global climate

change. Since the industrial revolution, CO_2 concentration in the atmosphere has increased from 280 ppmv (parts per million by volume) to 391 ppmv in 2011 [3]. In Europe, agricultural land use has been estimated to be the largest biospheric source of carbon emission, with a total carbon loss of 300 Mt C yr^{-1} (Mt C = million tons of carbon) [4].

As one of the main options to mitigate global climate change, carbon sequestration can remove CO_2 by transferring CO_2 from the atmosphere to the terrestrial biosphere [5, 6]. Terrestrial ecosystems can sequester CO_2 through photosynthesis and store or release carbon in four fundamental carbon pools (reservoirs with the capacity to store and release carbon) including aboveground biomass, belowground biomass, soil, and dead organic matter. Soil is the largest terrestrial carbon pool, which includes two major components: soil organic carbon (SOC) and soil inorganic carbon (SIC). However, most studies have been focused on SOC because SOC is the main component in most terrestrial ecosystems, and because SOC is the key factor of soil fertility and vegetation production [7]. According to the Intergovernmental Panel on Climate Change (IPCC) report, the terrestrial SOC contains about two times the amount of carbon stored in the atmosphere and vegetation [2].

Depending on land use and management practices, agricultural soils can act as a potential sink or source for atmospheric CO_2 [8–10]. Land conversions from natural to agricultural ecosystems can release large amounts of carbon [11]. It has been estimated that 50% of SOC in the top 20 cm depth of soil and 25–30% in the top 100 cm depth can be released following 30–50 years of land conversion to agriculture [12–14]. Agricultural cultivation of soil by plowing or other conventional tillage methods can also release CO_2 into the atmosphere, causing the decline of SOC pool [15]. With increasing demand for food and other living resources, agricultural intensification is generally seen as a necessary step to meet the joint food and environmental challenges [1]. Therefore, the way in which we design agricultural management practices has been considered as one of the most important strategies when trying to combat global climate change [14].

Recommended management practices (RMPs) are suggested as one of the principal ways in promoting SOC sequestration in agricultural soils [16]. By adopting RMPs, global SOC sequestration was estimated to vary from 0.4 to 0.8 Pg C yr^{-1}, which accounted for 33–100% of the total SOC sequestration potential in the world [17]. Some studies have reviewed the effects of different agricultural management practices on SOC stocks [8, 13, 16, 18–21]. Lal et al. [16] showed that if land management practice was designed properly, U.S. agricultural lands can be a major sink for carbon sequestration with the total carbon sequestration potential of 75–208 MMt C yr^{-1} (MMt C = million metric tons of carbon). West and Marland [20] used a full carbon cycle analysis (calculates both carbon input and carbon output) to compare carbon sequestration, carbon emissions, and net carbon flux associated with various tillage practices in the United States. VandenBygaart et al. [21] reviewed long-term studies in Canada to assess the influence of different management practices on SOC stocks. Estimates and uncertainties of the changes in SOC stock were compiled and utilized to estimate CO_2 emissions from agricultural soils around the world.

Among all RMPs, nitrogen (N) fertilization management and conservation tillage systems are two of the most highly recommended management practices in increasing SOC stocks in the agricultural soils, therefore having the potential to reduce the net CO_2 emissions into the atmosphere [22–25]. The major mechanism of N fertilizer addition in increasing SOC storage is through increases in crop yield and biomass production. In turn, more crop residues could be returned to the soil. In fact, the amount of crop residues returned to the soil is positively related to the amount of carbon sequestered [26–28]. Nonetheless, nitrogen fertilization's effect on SOC concentration varies among site-specific management, soil type, and climatic conditions [29, 30].

By definition, conservation tillage is any system that maintains at least 30% of crop residue on the soil surface with minimum or no tillage [31, 32]. The impact of various tillage systems on SOC content has been studied widely in field experiments. Lal et al. [9] reported a SOC sequestration potential of 24–40 Mt C yr^{-1} if adopting conservation tillage in the agricultural soils of the United States. No-till was estimated to emit less CO_2 (137 kg C ha^{-1}) than conventional tillage (168 kg C ha^{-1}), indicating that management practice from conventional tillage to no-till can enhance carbon sequestration [22]. No till often means more plant residue on the soil surface and less water and energy exchange between soil surface and the atmosphere. Hence, no till creates a system that favors SOC accumulation [33]. There are variations in the amount of carbon sequestration by no till practices due to differences in practice duration, climate conditions, soil types, crop rotation intensity, and management factors [32].

Several field studies have reported SOC stock changes as a result of nitrogen fertilization or conservation tillage management; however, data from such studies only provide site-specific examples of management impacts [30]. To gain a better understanding of management impacts on SOC stocks, a meta-analysis that compares and integrates the results from multiple studies is required [34]. There are several field studies that look at the effect of land management practices on SOC stocks since 2000. But there is no recent review after 2000. Therefore, a synthesis of studies published since 2000 will add new evidence to the effect of land management practices on SOC stocks. Furthermore, many of the recent reviews estimating carbon sequestration potential of cropland management practices have focused on European studies and little work has been done on the U.S. context. Another limitation in some of the reviews that examined management effects on SOC content is often based on studies that measure SOC stocks changes in the surface soil (<30 cm) [25]. For example, Baker et al. [35] criticized that higher SOC stocks as a result of no-till systems were almost always associated with soil samples collected above 30 cm. Therefore, a critical review that considers soil depth is very much needed to help validate no-till effects on SOC accumulation [36]. To my knowledge, such study is generally lacking.

The objective of this paper is to quantify the effects of N fertilization and conversion of management practice from conventional tillage (CT) to no till (NT) on soil organic carbon stocks in the United States. This will be accomplished by compiling available long-term experimental data from peer-reviewed journals. More specifically, the major goals of this review are twofold: (1) analyze the effects of N fertilization and tillage systems on SOC stocks, respectively, and (2) determine the main factors that can affect the response of SOC content to N fertilization and contrasting tillage systems.

2. Methods

2.1. Data sources and calculations

I used Google Scholar and Web of Science to search peer-reviewed literature between 2000 and 2014 with the keywords "nitrogen fertilization," "till or tillage," "soil organic carbon," "management practices* soil carbon." Studies on the effect of nitrogen fertilization and tillage systems on SOC stocks from literature search were filtered to include only studies carried out in the agricultural soils of the conterminous United States. Any study included in the analysis had to meet the following criteria: (1) experiment set-up in the field had to be clearly stated, including the start and end dates of the study or duration of the treatment, soil sampling depth, the amount of nitrogen fertilizer applied in the field over time, tillage system used, etc. (2) SOC stocks per unit area or SOC concentrations and soil bulk density had to be reported. (3) Changes in SOC stocks or SOC concentrations and soil bulk density had to be attributed to different nitrogen application rates or to contrasting tillage systems. (4) No crop residue removal should have occurred over the study period.

Data from reviewed papers were extracted. For fertilizer and tillage experiments, a control treatment is contrasted with an alternative treatment. For fertilizer experiments, I compared unfertilized (control) treatment with fertilized. There were 145 paired comparisons of measured SOC stocks between fertilized and control treatments. For tillage experiments, there were a total of 187 paired comparisons with contrasting tillage system: no tillage management was practiced in 186 paired-experiments, conventional tillage was practiced in 187 paired-experiments, and reduced till was applied in 38 paired-experiments. The key independent variable, total nitrogen applied, was calculated by adding up the amount of nitrogen fertilizer applied each year over the study period. For some studies, nitrogen application was not applied at a constant rate; then, the total amount of nitrogen fertilizer applied was calculated by adding up the actual application rate across the duration of experiment. Otherwise, total nitrogen applied was calculated by multiplying nitrogen fertilizer rate per year with treatment durations. The response variable, paired log difference of SOC measurement between fertilized and unfertilized/control practices, was calculated using Eq. (1), and was used to eliminate the differences of means and variances among different studies. In this particular case, if the response ratio is greater than zero, management practice from fertilized to unfertilized treatment increases SOC stocks.

$$\ln\ (\text{fertilized}) - \ln\ (\text{control}) = \ln\ (\text{fertilized}/\text{control}) \tag{1}$$

For tillage analysis, three principal tillage systems were considered: conventional tillage (CT), reduced till (RT), and no till (NT) (in some studies, no till treatment was set up as conservation tillage). The response variables, paired log difference between no till and conventional till, no till and reduced till, reduced till and conventional till, were calculated using Eqs. (2)–(4). Here, if the response ratio, ln (NT/CT), is greater than zero, no till is said to increase SOC stocks compared with conventional tilled system. Similarly, if response ratios, ln (NT/RT) and ln (RT/CT), are greater than zero, SOC stocks increase when changing from no till to reduced till, reduced till to conventional till, respectively.

$$\ln\ (\mathrm{NT}) - \ln\ (\mathrm{CT}) = \ln\ (\mathrm{NT}/\mathrm{CT}) \tag{2}$$

$$\ln\ (\mathrm{NT}) - \ln\ (\mathrm{RT}) = \ln\ (\mathrm{NT}/\mathrm{RT}) \tag{3}$$

$$\ln\ (\mathrm{RT}) - \ln\ (\mathrm{CT}) = \ln\ (\mathrm{RT}/\mathrm{CT}) \tag{4}$$

Some studies in my dataset only reported SOC concentration and soil bulk density instead of total SOC stock. In that case, SOC stock was calculated as follows (Eq. (5)):

$$\mathrm{SOC}\ \left(\mathrm{Mg\ C\ ha^{-1}}\right) = \mathrm{SOC}\ (\%)^{*}\ \text{Soil bulk density}\ \left(\mathrm{Mg/m^{3}}\right)^{*}\ \text{Soil sampling depth (cm)} \tag{5}$$

The following were considered environmental and edaphic variables: experimental site, treatment duration (time since practice), crop rotation system, cropping index, soil sampling depth, soil texture, mean annual temperature (MAT), and mean annual precipitation (MAP). Based on crop rotation system, a discrete cropping index was calculated by incorporating the number of crops rotated per year, and the percentage of corn in the cropping system (after Alvarez [30]). The calculation of cropping index was also based on two assumptions: (1) residue produced from corn was twice as much as from other crops, and (2) two crops per year produced twice the amount of residue of one crop per year [24, 30]. Soil texture was categorized into three types: fine, loamy, and coarse. In terms of climatic data, MAT and MAP were extracted from the reviewed papers. If for any reason, MAT and MAP were not reported or missing from the study, they were estimated from the following website: http://www.ncdc.noaa.gov/.

2.2. Statistical analysis

2.2.1. Analysis of nitrogen fertilization and SOC stocks

First, a paired t-test was used to test whether SOC stock with fertilizer is significantly different from SOC without fertilizer (control). This was done by testing changes in measured SOC between fertilized and control treatments against zero at a significance level of 0.05. Then, bivariate and multivariate regression models were developed to investigate the relationship between paired log difference of measured SOC stocks (ln (fertilized/control)) and total nitrogen applied in a context shaped by variables that can moderate the effect of fertilization on SOC stock. Here, experimental location as random effect was combined with multivariate regression model because more than one measurement was taken from the same geographic location. Location as random effect relaxes the assumption that data of different plots with alternative treatments taken from the same site are independent from each other. Variables considered in the model include treatment duration, cropping index, soil sampling depth, soil texture, mean annual temperature, and mean annual precipitation. Finally, paired log difference of measured SOC stocks between fertilized and control treatment was further analyzed for the effects of relevant environmental and edaphic variables (e.g., soil texture). Means and 95% confidence intervals (CIs) of paired log difference in measured SOC across the dataset were reported. If the 95% CIs of paired log difference in measured SOC stocks for a given variable does not overlap with zero, the response of that variable to fertilizer effect is said to be significantly different from the control [34].

2.2.2. *Analysis of tillage systems and SOC stocks*

Paired t-tests were first conducted to compare the effects of contrasting tillage systems (no till vs. conventional till, no till vs. reduced till, and reduced till vs. conventional till) on SOC stocks at a significance level of 0.05. Linear and curvilinear models were tested to see which model fits better with the dataset. Linear regression model was therefore chosen for this analysis. I estimated the correlation between treatment duration and paired log difference in measured SOC produced from fertilizer. Multivariate regression model was also applied to develop equations that explain the effects of no tillage system on SOC stocks with control variables that can potentially affect its response. Location as random effect model was also incorporated in the multivariate regression model with the same process that was applied in fertilizer experiments. Lastly, the effects of relevant environmental and edaphic variables on paired log difference of measured SOC between no till and conventional till was further analyzed with the mean and 95% CIs calculated. All statistical analyses were performed in the Stata software package (StataCorp. 2013. Stata Statistical Software: Release 13. College Station, TX: StataCorp LP).

3. Results and discussion

3.1. Analysis of nitrogen fertilization and SOC stocks

A total of 145 paired experiments with varying nitrogen fertilization rates were compiled in the database for this analysis (**Table 1**). The database covers 10 states. Of all the 145-paired studies, the total nitrogen fertilization applied varied from 0.089 to 6.44 Mg N ha^{-1}. Changes of SOC stock produced from varying nitrogen fertilization treatments were between −14 and 22 Mg C ha^{-1}, with an average of 2.32 Mg C ha^{-1}. The treatment durations of these experiments were between 2 and 27 years, with an average of 10.8 years. The soil sampling depth spanned a wide range from 7.6 to 120 cm, with an average of 48.4 cm. In terms of weather attributes, the lowest and highest mean annual temperatures were 7 and 17°C, averaging 11.4°C. The mean annual precipitation ranged from 357 to 1400 mm at an average of 762.7 mm.

A paired t-test showed that measured SOC under fertilizer treatments was significantly different ($p < 0.001$, $t = 5.74$, degrees of freedom = 144) from measured SOC under control treatments.

Description	Mean	Std	Min	Max	# of observations
Total nitrogen applied (Mg N ha^{-1})	0.99	0.89	0.089	6.44	145
ln (fertilized/control)	0.03	0.06	−0.165	0.28	145
Treatment duration (years)	10.8	6.6	2	27	145
Soil sampling depth (cm)	48.4	35.7	7.6	120	145
Mean annual temperature (°C)	11.4	3.4	7	17	145
Mean annual precipitation (mm)	763	300	357	1400	145

Table 1. Summary statistics for the paired data of N fertilizer experiments used in this study.

Roughly 66% (n = 95) of the total observations showed the positive effect of nitrogen fertilizer on SOC storage with no crop residue removal, whereas about 32% (n = 47) and 2% (n = 3) showed negative and no correlation between total nitrogen fertilization and SOC content, respectively. When total nitrogen fertilization rate is higher than 3.51 Mg ha^{-1}, no SOC depletion occurred (**Figure 1**). Total nitrogen applied had a significant positive impact on measured SOC change between fertilized and control treatment ($p < 0.001$) (**Figure 1**). As the application of total nitrogen fertilizer increased, the paired log differences of measured SOC stocks between fertilized and control treatments increased. Specifically, when total nitrogen fertilizer increased by 1 Mg ha^{-1}, the paired log differences of measured SOC stocks increased by 0.02. Measured SOC stock increased by 2% relative to control treatment.

The increases in SOC level as a result of N fertilizer addition are attributable to the increases in net primary productivity and residue-C input [37]. A strong negative correlation between SOC content and crop residue production was observed under N deficit by Campbell and Zentner [38, 39]. The significant positive effect of N additions on SOC level detected agrees with a review by West and Post [22] based on a compiled global database of 67 long-term agricultural experiments. However, the magnitude of this effect varied from significant increase [40–45] to only mild increase in the level of SOC [46–49].

In this study, the effect of N additions on measured SOC stocks was, however, moderated by the relevant environmental and edaphic characteristics, including cropping index, soil sampling depth, soil texture index, mean annual temperature (MAT), and mean annual precipitation (MAP). Here, a multivariate regression model (Eq. (6)) with random effect was developed to characterize the relationship between paired log differences of measured SOC stocks (ln (fertilized/control)) and the total nitrogen applied in the experiment (**Table 2**).

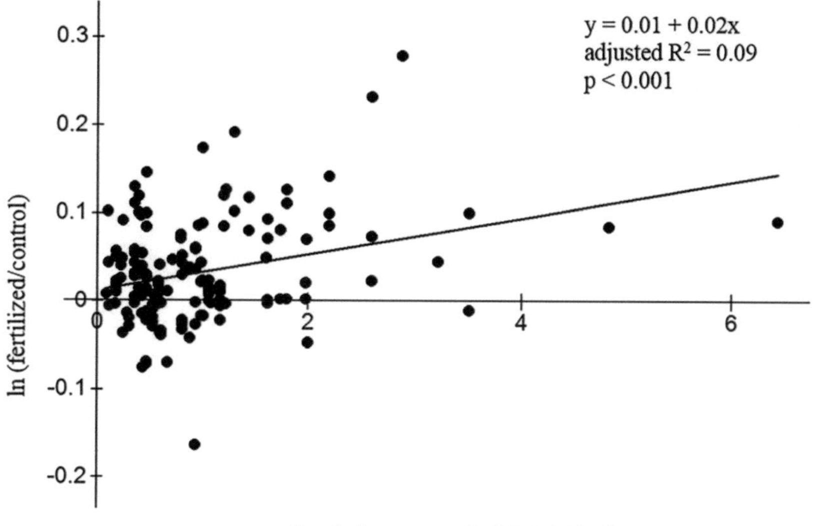

Figure 1. Paired log difference of soil organic carbon between fertilized and control measurement (ln (fertilized/control)) plotted against the total nitrogen applied in experiments with no crop residue removal.

Dependent variable	ln (fertilized/control)
Total nitrogen applied	0.013** [2.120]
Cropping index	0.063*** [4.513]
Soil sampling depth	−0.000 [−0.348]
Soil texture index	0.009 [0.700]
MAT (mean annual temperature)	−0.003 [−1.023]
MAP (mean annual precipitation)	−0.000 [−0.950]
Constant	−0.018 [−0.571]
Observations	145
Number of location	12

Table 2. This table presents multivariate regression results for relationship between paired log difference of soil organic carbon between fertilized and control measurement and the total nitrogen applied in experiments with no crop residue removal. The dependent variable is calculated as follows: ln (fertilized/control) = ln (fertilized) − ln (control). Independent variables are total nitrogen applied (Mg N ha^{-1}), cropping index, soil sampling depth, soil texture index (1 = fine, 2 = loamy, 3 = coarse), and climate condition including MAT (°C) (mean annual temperature) and MAP (mm) (mean annual precipitation). The t-values are given in brackets. ***, **, and * denote significance at the 0.01, 0.05, and 0.1 level, respectively.

$$\ln\ (\text{fertilized}/\text{control}) = -0.018 + 0.013\ \text{N} + 0.063\ \text{C}_\text{i} - 0.000\ \text{D} + 0.009\ \text{SI} - 0.003\ \text{T} - 0.000\ \text{P} \tag{6}$$

($p < 0.01$, number of observations = 145, number of locations = 12).

where N is the total nitrogen applied (Mg N ha^{-1}), C_i is the cropping index, D is the soil sampling depth (cm), SI is the soil texture index, T is the mean annual temperature (°C), and P is the mean annual precipitation (mm).

Cropping system significantly increased paired log difference of measured SOC stocks between fertilized and control treatment ($p < 0.01$; **Table 2**). When cropping index increased by 1, paired log difference of measured SOC stocks increased by 0.063. Measured SOC stock under fertilized treatment was 6.5% higher than SOC stock under control treatment. Increases in cropping index can be achieved by either rotating more crops per year or incorporating corn as the main component in the cropping system. By rotating more crops per year, net primary productivity of the cropland increased. Hence, SOC storage increased, therefore contributing to the absorption of the atmospheric carbon dioxide. Due to a large expansion in ethanol production in the United States, the market price of corn has experienced significant overall increases in recent years. Response to high corn prices, farmers increasingly choose to increase corn acreage at the expense of other crops, such as soybean. Therefore, due to reduced soybean production, soybean price also increases significantly in recent years. These socioeconomic factors have induced a series of cropping system changes in the Midwest Corn Belt. Hence, cropping systems in the Midwest Corn Belt include three major types: continuous corn cropping, continuous soybean cropping, and corn-soybean/soybean-corn rotation.

SOC level under continuous corn is often higher than under corn-soybean/soybean-corn rotation because corn produces more biomass than soybean does [18, 37]. Measured SOC stocks increased significantly ($p < 0.001$, $t = 9.41$, degrees of freedom = 45) when cropping index equals 2 (**Figure 2**). More specifically, measured SOC stock with fertilizer increased by 18% compared with control treatment. One of the possible cropping system when the cropping index equals 2, is continuous corn cropping. West and Post [22] found that as rotation intensity increased, SOC sequestration rate increased by 200 ± 120 kg C ha^{-1} yr^{-1}, with an exception of change from continuous corn to corn-soybean/soybean-corn rotation. However, when cropping index is lower than 2, the effect of cropping index on changes of measured SOC stocks was not significantly different among cropping sequences.

There was no significant correlation between soil sampling depth and changes in measured SOC stocks produced by nitrogen fertilization application (**Table 2**). SOC stocks significantly increased among all soil sampling depths measured in 145-paired experiments (**Figure 3**). Means and 95% CIs of paired log differences of measured SOC stocks overlapped, which means that the N effect on SOC stocks did not differ across all sampling depths. Measured SOC increased profoundly (10%) when soil sampling depth was below 30 cm.

Soil texture includes three categories: fine (=1), loamy (=2), and coarse (=3). Here, the correlation between soil texture index and paired log difference of measured SOC stocks was not detected (**Table 2**). This contradicts with Alvarez [30] who observed a significant positive

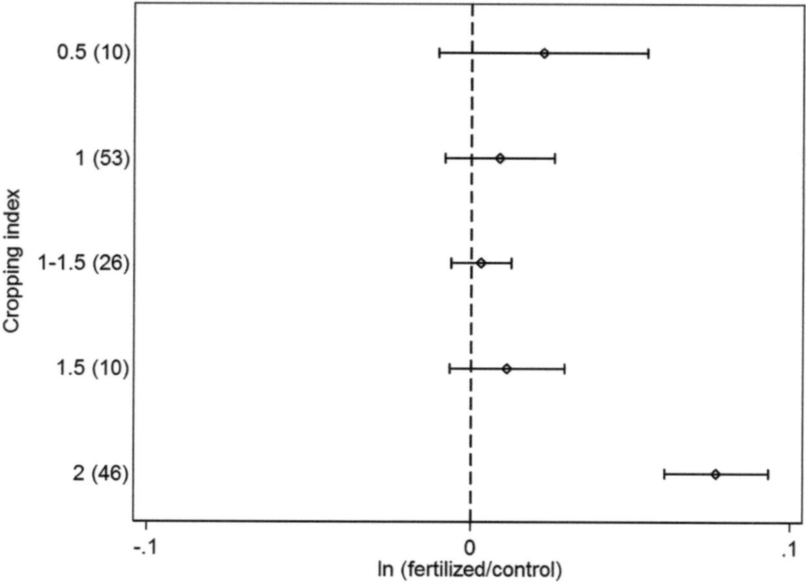

Figure 2. The effects of cropping index on paired log difference of soil organic carbon between fertilized and control measurement (95% confidence intervals are shown and numbers of observations are included in parentheses).

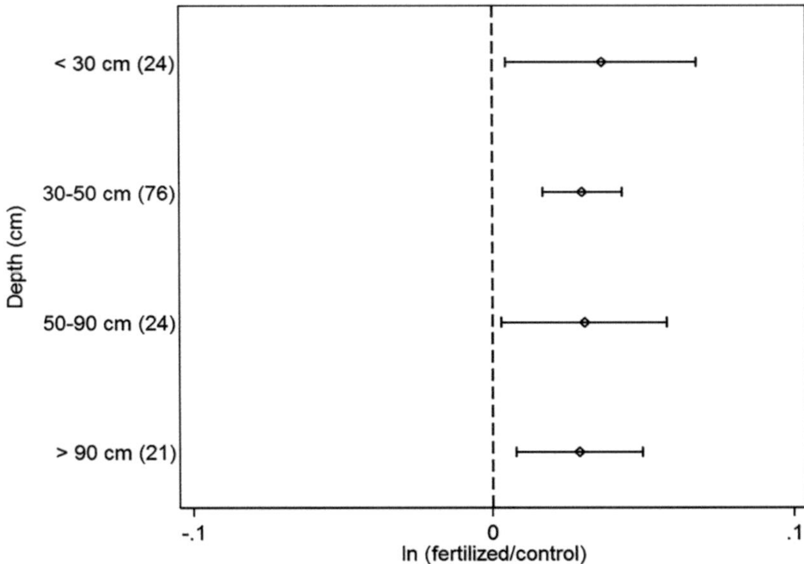

Figure 3. Paired log difference of soil organic carbon between fertilized and control measurement across the soil sampling depth (95% confidence intervals are shown and numbers of observations are included in parentheses).

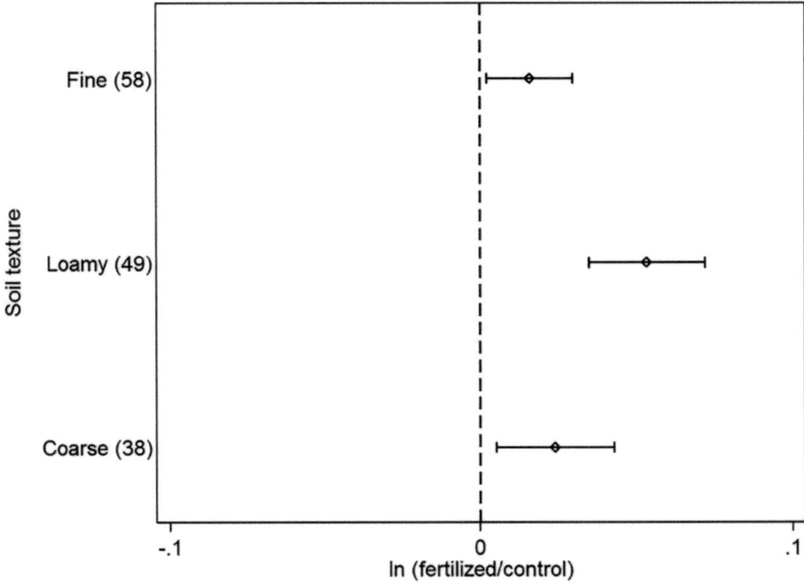

Figure 4. The effects of soil texture on paired log difference of soil organic carbon between fertilized and control measurement (95% confidence intervals are shown and numbers of observations are included in parentheses).

relationship between soil texture index and changes in SOC stocks. Coarse-textured soils are often associated with lower soil fertility; therefore, might response stronger to nitrogen fertilizer addition if other factors are held constant [30]. The effect of nitrogen fertilizer on SOC stocks was significant across all soil types in this analysis (**Figure 4**). In fact, areas with fine- and coarse-textured soils did not differ significantly in terms of their effects on measured SOC stocks. On average, soils with loamy texture significantly increased ($p < 0.0001$, $t = 5.71$, degrees of freedom = 48) SOC stocks by 13% than those of fine- (4.6%) and coarse-textured (6.9%) soils.

In terms of climates, no statistically significant correlation was found between mean annual temperature and paired log difference of measured SOC, mean annual precipitation and paired log difference of measured SOC, respectively (**Table 2**). Previous studies that examined relationship between climate conditions and N effect have come to mix conclusions. Parton et al. [50] reported that temperature can negatively affect residue-C transition to SOC stocks. Therefore, it is expected that the effect of nitrogen fertilization on SOC stocks is greater in temperate climates compared with tropical climates [30].

Furthermore, I found that areas with temperature < 12°C sequestered significantly more SOC, but not in areas with temperature ranging from 12 to 15°C (**Figure 5**). In fact, the highest SOC increase located in areas with temperature lower than 8°C (+15%). Increase in measured SOC was also significant in areas with temperature above 15°C. Considering distributions of measured SOC difference across areas with various mean annual precipitations, N fertilizer had a significant

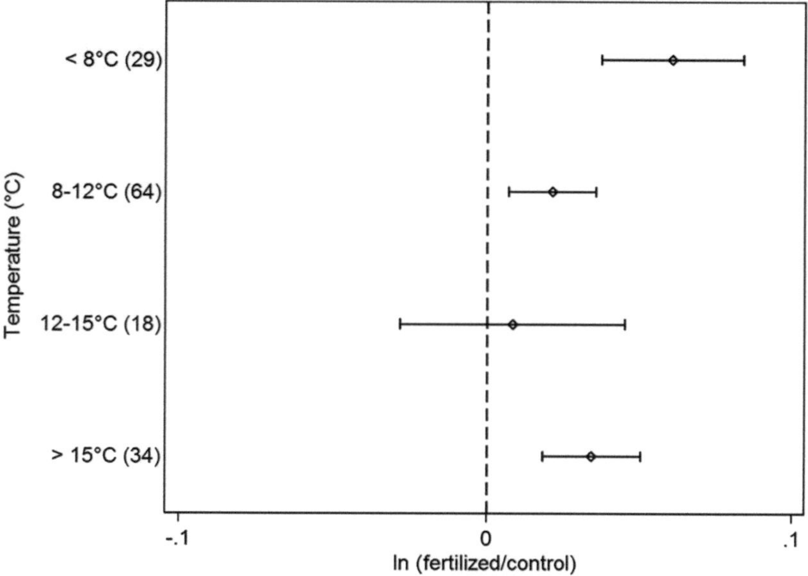

Figure 5. The effects of mean annual temperature on paired log difference of soil organic carbon between fertilized and control (95% confidence intervals are shown and numbers of observations are included in parentheses).

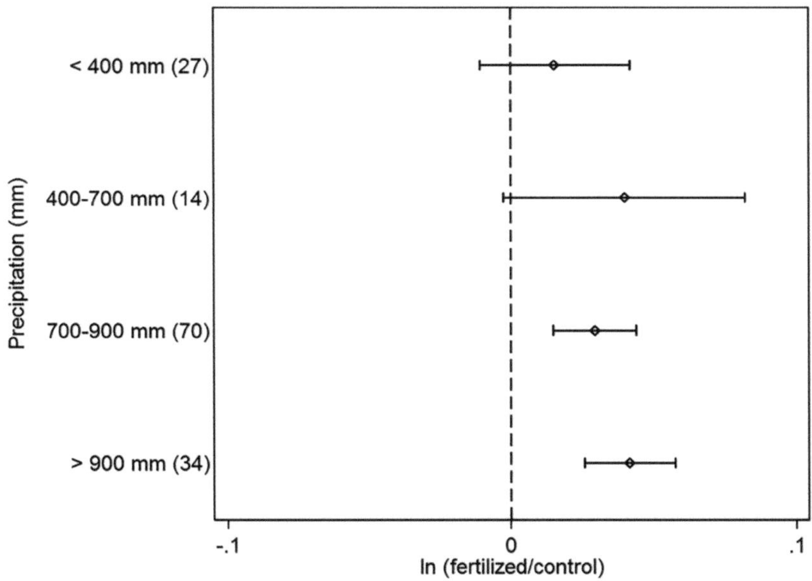

Figure 6. The effects of mean annual precipitation on paired log difference of soil organic carbon between fertilized and control measurement (95% confidence intervals are shown and numbers of observations are included in parentheses).

impact on SOC stocks in the higher rainfall (>700 mm) areas, but had no effect on SOC stocks in lower rainfall (<700 mm) areas (**Figure 6**).

Even though nitrogen fertilization can result in SOC sequestration, its potential to remove carbon from the atmosphere is still debatable and requires a comprehensive evaluation of the whole process from fertilizer manufacture to transportation, and finally to applications in the fields [22]. The production of N fertilizers involves energy input from fossil fuel combustion, which in turn leads to carbon emissions back into the atmosphere. There are also post-production carbon emissions from fertilizer packaging, transportation, and field application [51]. Average carbon emissions associated with the production and use of N fertilizers were estimated to be 1.2 Mg C Mg^{-1} N applied [22, 52]. In conclusion, to evaluate carbon mitigation potential of N fertilization management, a comprehensive assessment from N manufacture, delivery, to application is required.

3.2. Analysis of tillage systems and SOC stocks

Three tillage systems were considered in this analysis, which include no till (NT), reduced till (RT), and conventional till (CT). Overall, studies compiled in this database comprise 187-paired experiments. Of all 187 paired data, 186 cases (99%) report changes in SOC stocks between no till and conventional till, 37 cases (20%) measure ΔSOC stocks between no till and reduced till, and 38 cases (20%) for SOC stocks changes from reduced till to conventional till (**Table 3**). The database covers 20 states. Among all 186 paired comparisons, paired log difference of measured SOC from conventional tillage to no tillage ranged from −0.37 to 0.6, with an average of 0.089. In other words, changes in measured SOC with no till management

Description	Mean	Std	Min	Max	# of observations
ln (NT/CT)	0.089	0.151	−0.366	0.6	186
ln (NT/RT)	0.007	0.147	−0.213	0.569	37
ln (RT/CT)	0.065	0.083	−0.107	0.272	38
Treatment duration (years)	12.84	10.19	2	45	187
Soil sampling depth (cm)	35.75	27.53	6	150	187
Mean annual temperature (°C)	13.3	4.43	5.5	23.5	187
Mean annual precipitation (mm)	945	324	305	1584	187

Table 3. Summary statistics for the paired data of tillage experiments used in this study.

ranged from −31% to +82% compared with conventional tillage system. Paired log difference of measured SOC from reduced tillage to no tillage varied from −0.21 to 0.57, with a mean of 0.007. This suggests that changes in measured SOC with no till practice can decrease up to 19% and increase as much as 77% relative to reduced tillage. Paired log difference of measured SOC content from conventional tillage to reduced tillage ranged from −0.11 to 0.27, averaging around 0.06. The differences of measured SOC level between reduced till and conventional till varied from −11% to +31%. Of all 187 paired comparisons, the treatment durations were from 2 to 45 years, with an average of 12.8 years. Soil depth sampled was in a range of 6–150 cm, with an average of 35.7 cm. Mean annual temperature was from 5.5 to 23.5°C at an average of 13.3°C, and mean annual precipitation ranged from 305 to 1584 mm, averaging 945 mm.

Of all 186 observations that measured changes in SOC storage between no till and conventional till, approximately 71% ($n = 133$) of the total observations, showed positive values. Of all 37 paired experiments that reported SOC differences between no till and reduced till, more than half of the total cases (57%; $n = 21$) showed negative results, with 16 (43%) cases showed positive values. In contrast, among all 38 studies that reported changes in measured SOC stocks from conventional till to reduced till, only 2 cases showed negative values, 6 cases were no change, and the remaining 30 cases (79%) were positive values. Paired t-tests showed significant differences in measured SOC stocks between no till and conventional till ($p < 0.001$, $t = 8.06$, degrees of freedom = 185), reduced till and conventional till ($p < 0.001$, $t = 4.83$, degrees of freedom = 37), respectively. SOC stocks under no till and reduced till were on average 9% and 7% greater than those of conventional till. However, paired t-tests showed no significant differences between no till and reduced till. This could be true or it could be due to the low number of observations for this measure.

No significant correlation between paired log difference of measured SOC content and duration time was detected between reduced till and conventional till. However, paired log difference of measured SOC between no till and reduced till was significantly dependent on time since management practice ($p < 0.001$; **Figure 7**). Again, due to its low number of observations ($n = 37$), I won't further analyze this measure in this study. As expected, the differences of measured SOC stock between no till and conventional till were also significantly dependent on length of time since conversion ($p < 0.001$; **Figure 7**). The longer the time in no till management, the greater the amount of SOC stocks compared to conventional tilled fields. More specifically,

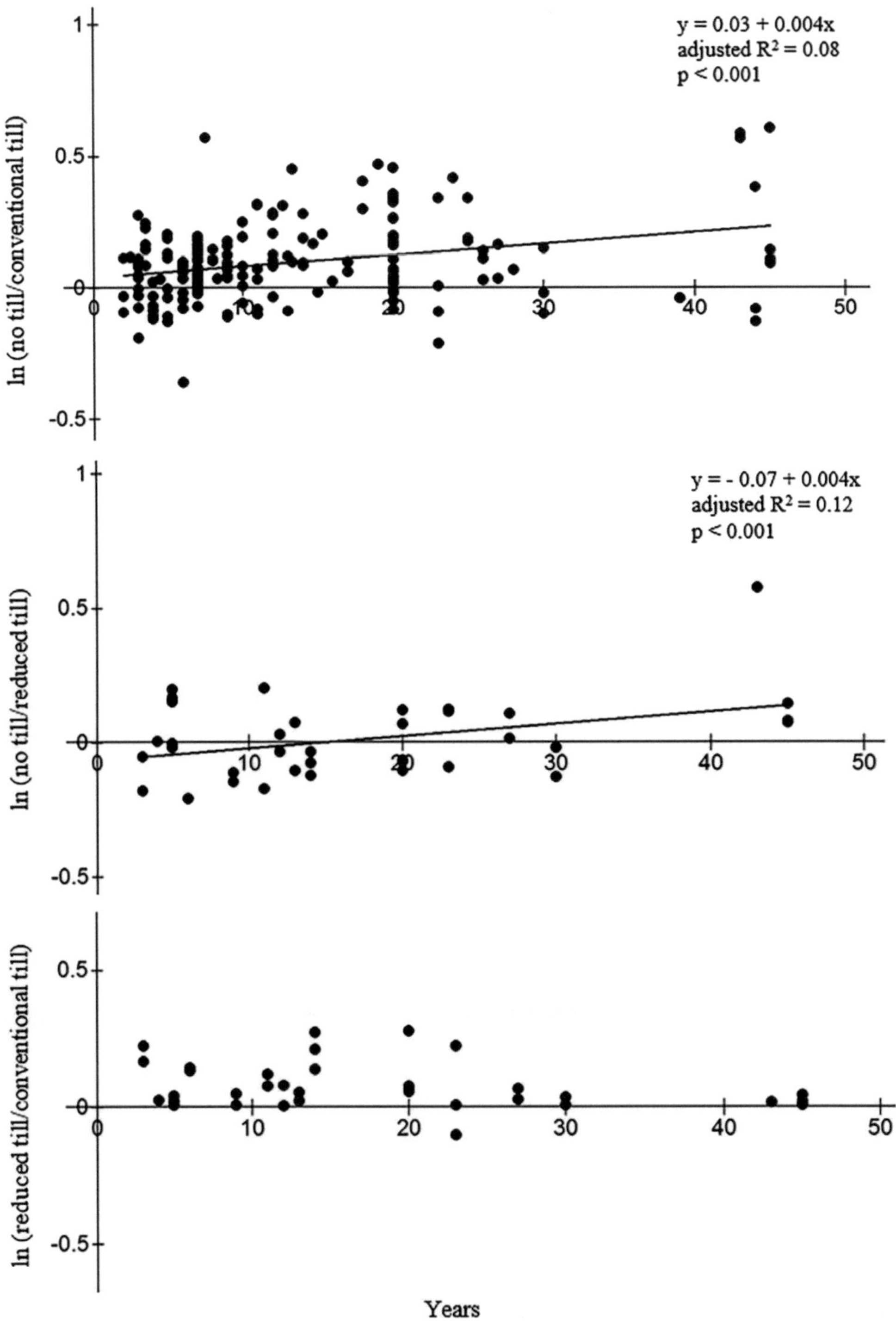

Figure 7. Paired log difference of soil organic carbon between contrasting tillage systems plotted against treatment durations. Here, tillage systems include no-till, conventional till, and reduced till.

if treatment duration increases by 1 year, SOC stock would increase by 0.4% when changing from conventional tillage to no tillage system.

Increases in measured SOC stocks occurred in the soil when the duration of no tillage treatment was beyond 5 years (**Figure 8**). This result is consistent with the findings summarized by West and Post [22], that there was a delayed response of no till management on SOC stocks with peak sequestration rates in 5–10 years. Despite the high degree of variations in climate conditions, soil types, cropping systems, and other associated site characteristics, differences between conventional till and no till were still significantly ($p < 0.05$) time-dependent: SOC stock increased as the time in no-till management increased. A multivariate regression with random effects model (Eq. (7)) was established to account for the associated environmental and edaphic characteristics (**Table 4**).

$$\begin{aligned} \ln\ (\text{no till/conventional till}) = & -0.157 + 0.004\ \text{DT} + 0.009\ \text{C}_\text{i} \\ & +0.0001\ \text{D} + 0.038\ \text{SI} + 0.003\ \text{T} + 0.0001\ \text{P} \end{aligned} \tag{7}$$

($p < 0.05$, number of observations = 186, number of locations = 70).

where DT is the treatment duration time (years), C_i is the cropping index, D is the soil sampling depth (cm), SI is the soil texture index, T is the mean annual temperature (°C), and P is the mean annual precipitation (mm).

There was no significant correlation between cropping system and changes in measured SOC stocks (**Table 4**). So did soil texture, mean annual temperature, and mean annual precipitation. Increases in measured SOC stocks occurred significantly when cropping index was greater than

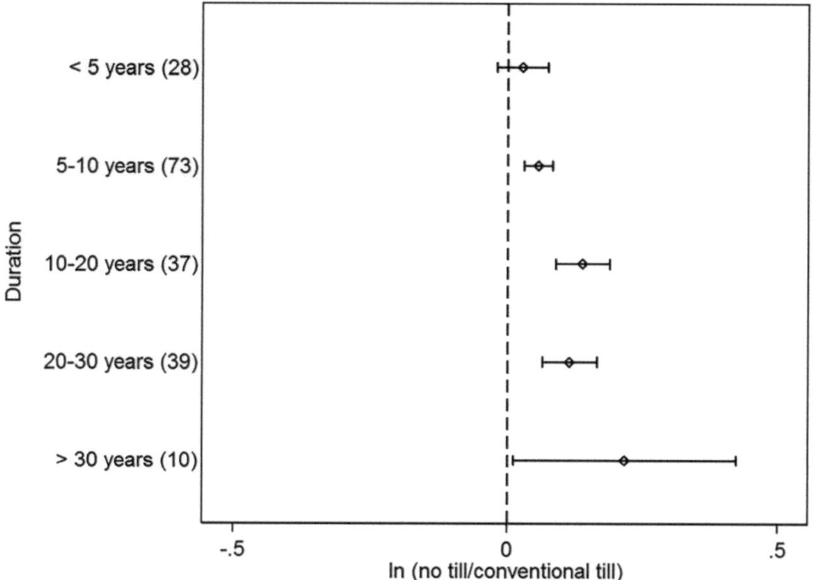

Figure 8. The effects of treatment durations on paired log difference of soil organic carbon measurement between no till and conventional till (95% confidence intervals are shown and numbers of observations are included in parentheses).

Dependent variable	ln (no till/conventional till)
Duration time	0.004** [2.431]
Cropping index	0.009 [1.140]
Soil sampling depth	0.0001 [0.228]
Soil texture index	0.038 [1.316]
MAT (mean annual temperature)	0.003 [0.703]
MAP (mean annual precipitation)	0.0001 [0.877]
Constant	−0.157* [−1.850]
Observations	186
Number of location	70

Table 4. This table presents multivariate regression results for relationship between paired log difference of soil organic carbon measurement from conventional till to no till and treatment duration. The dependent variable is calculated as follows: ln (no till/conventional till) = ln (no till) − ln (conventional till). Independent variables are treatment duration, cropping index, soil sampling depth, soil texture index (1 = fine, 2 = loamy, 3 = coarse), and climate condition including MAT (°C) (mean annual temperature) and MAP (mm) (mean annual precipitation). The t-values are given in brackets. ***, **, and * denote significance at the 0.01, 0.05, and 0.1 level, respectively.

0.5 (**Figure 9**). However, paired log differences of measured SOC between no till and conventional till were not significantly different from each other when cropping index was greater than 0.5. On average, SOC content from conventional tillage to no tillage increased by roughly 9% across all cropping sequences when cropping index was greater than 0.5. All three types of soil texture (fine, loamy, and coarse) had significant effects on measured SOC change from conventional till to no till (**Figure 10**). The conversion from conventional tillage to no tillage system had no effect on changes of measured SOC stocks in the lower rainfall (<900 mm) areas, but significantly increased measured SOC stocks in higher rainfall areas (>900 mm) (**Figure 11**).

There was no significant association between soil sampling depth and paired log difference of measured SOC stocks (**Table 4**). However, the distribution of paired log difference of measured SOC across all soil sampling depths showed significant ($p < 0.001$) increases in SOC content in the surface soil (<50 cm) and above 90 cm. In particular, increases in measured SOC stocks were greater (+34%) in the upper 30 cm of the soil profile relative to 30–50 cm of the soil profile. This result is consistent with previous studies.

Considering the distribution of paired log difference of measured SOC content across areas with different mean annual temperatures, there was no significant change in measured SOC level between no till and conventional till in areas with low temperature (<8°C). Areas with temperature above 8°C can significantly increase measured SOC stocks when applying no tillage system (**Figure 12**).

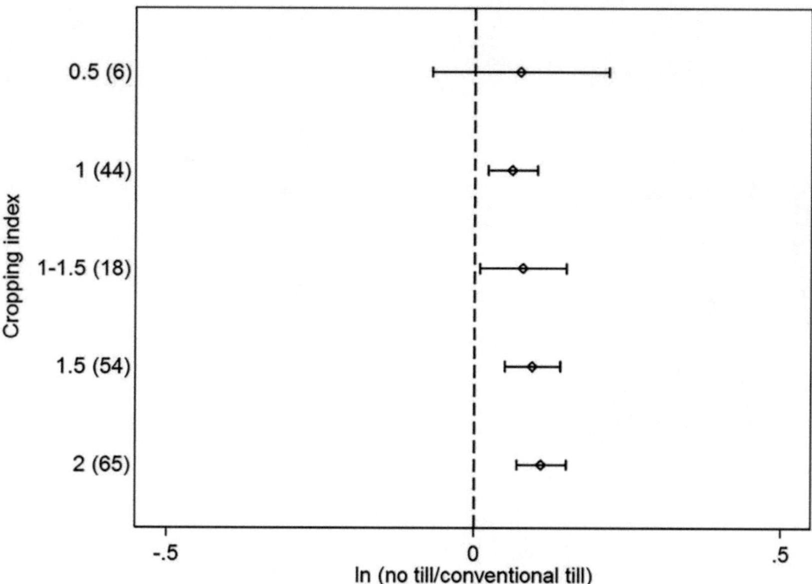

Figure 9. The effects of cropping index on paired log difference of soil organic carbon measurement between no till and conventional till (95% confidence intervals are shown and numbers of observations are included in parentheses).

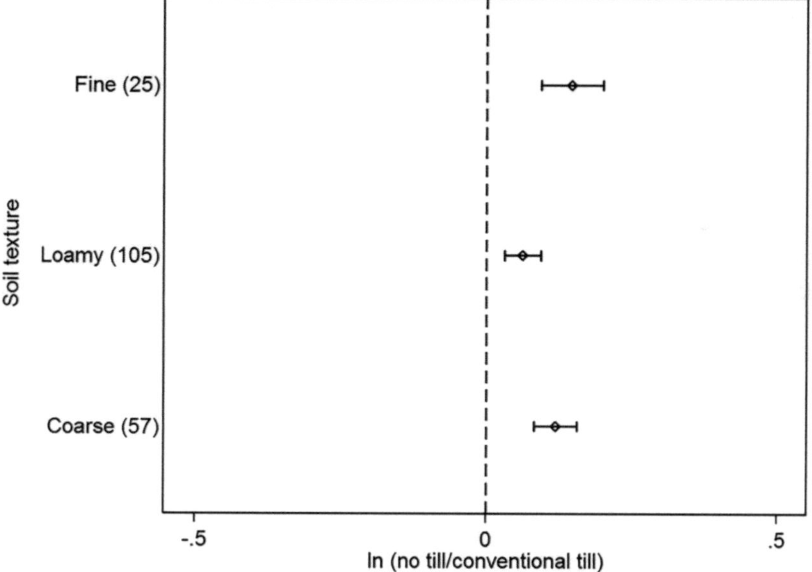

Figure 10. The effects of soil texture on paired log difference of soil organic carbon measurement between no till and conventional till (95% confidence intervals are shown and numbers of observations are included in parentheses).

Management practice from conventional tillage to conservation tillage is found to increase SOC levels; however, this is not always effective, especially in fine-textured and poorly drained soils and cold weather conditions [53–56]. Moreover, it is possible that no till or conservation till could

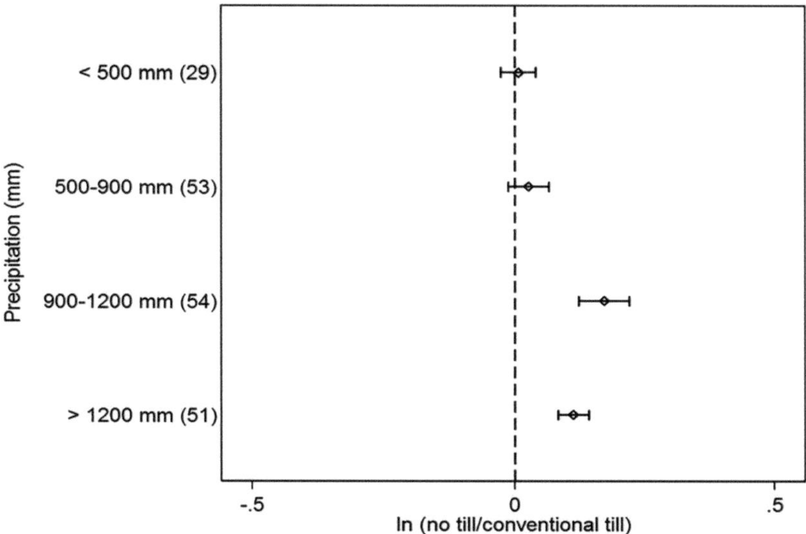

Figure 11. The effects of mean annual precipitation on paired log difference of soil organic carbon measurement between no till and conventional till (95% confidence intervals are shown and numbers of observations are included in parentheses).

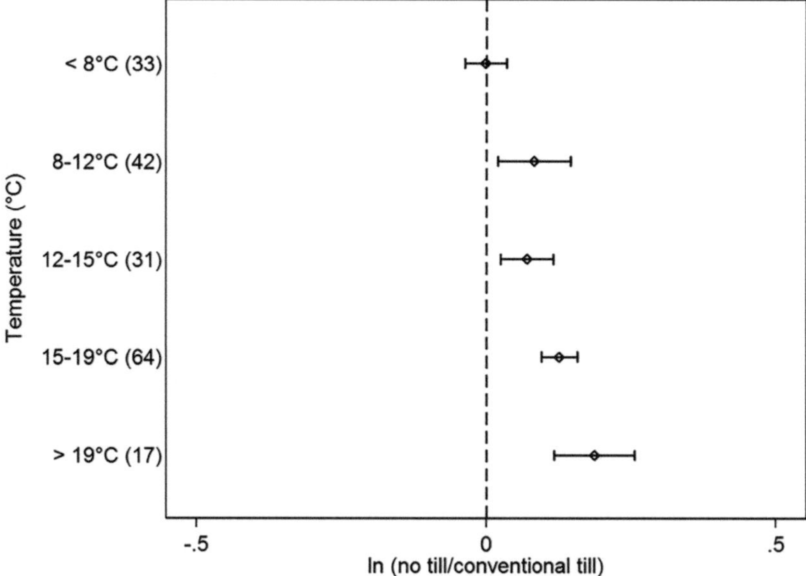

Figure 12. The effects of mean annual temperature on paired log difference of soil organic carbon measurement between no till and conventional till (95% confidence intervals are shown and numbers of observations are included in parentheses).

contribute to N_2O emissions, another GHG with even stronger climate warming potential [57–59]. The estimated N_2O emissions as a result of no till management are varied and inconsistent: some reported positive impacts, whereas some reported negative or no measurable impacts on

N_2O emissions [60]. The N_2O emissions may counterbalance all or some of the increased SOC content in terms of GHG mitigation potential in agriculture [61]. Therefore, to assess the capability of conservation tillage systems in mitigating global climate change, a systematic evaluation of all GHG emissions should be considered. Nonetheless, conservation tillage systems are a viable option that can sequester CO_2 from the atmosphere.

4. Conclusion

This review quantitatively evaluated the impacts of nitrogen fertilization management and conservation tillage systems on SOC stocks in the agricultural soils of the United States. The results presented here showed that N fertilizer additions had significant positive impact on SOC content, but the magnitude of this effect varied. In fact, the effect of N fertilization treatment on SOC stocks was moderated by cropping rotation system. As the cropping intensity increased, measured SOC content under fertilized treatment also increased. Soil texture and climate conditions, including mean annual temperature and mean annual precipitation, did not have significant impacts on differences of measured SOC stocks between fertilized and control treatments.

Significant differences in SOC stocks were found between no till and conventional till, as well as between no till and reduced till. However, SOC stocks between no till and reduced till were not significantly different. Differences of SOC content due to management changes from conventional tillage to no till system were significantly larger when treatment duration was longer. This study also showed a delayed response of SOC level to no till management with increases in measured SOC occurring beyond 5 years. Crop rotation system, soil texture, mean annual temperature, and mean annual precipitation did not have significant effects on SOC stocks. To summarize, paired log differences of measured SOC content from conventional tillage to conservation tillage were only significantly dependent on time since management.

To help combat global climate change, it is of great importance to identify changes in land management practices that can promote carbon sequestration and mitigate the enhanced greenhouse gas effect. The study recorded the responses of SOC stocks to changes in management practices and confirmed that adoption of N fertilizer additions and conservation tillage systems can contribute to increased SOC stocks in the agricultural soils of the United States. However, the evaluation of net carbon dioxide mitigation potential of these two recommended management practices should be carried out using a full carbon and greenhouse gas accounting method, which comprehensively considers both carbon input and carbon output to the agricultural systems. To conclude, agricultural soils can act as an important carbon sink to offset atmospheric CO_2 emissions when management practices are designed appropriately, as well as with proper incentives and technological advancements. Confidence intervals for estimates of carbon sequestration rates in this study can be incorporated in policy and carbon cycle modeling analysis to provide more accurate estimates of C sequestration potential at regional and global scales.

Author details

Meimei Lin

Address all correspondence to: meimei.lin@armstrong.edu

Department of Geology and Geography, Georgia Southern University, GA, USA

References

[1] Foley JA, DeFries RS, Asner G, Barford CC, Bonan G, Carpenter SR, Chapin F, Coe M, Daily G, Gibbs HK, Helkowski J, Holloway T, Howard E, Kucharik CJ, Monfreda C, Patz J, Prentice IC, Ramankutty N, Snyder PK. Global consequences of land use. Science. 2005;**309**:570-574

[2] IPCC. Intergovernmental Panel on Climate Change, Land Use, Land-Use Change, and Forestry. Cambridge: Cambridge University Press; 2000

[3] IPCC. Mitigation of Climate Change of Working Groups III to the fifth assessment Report of the Intergovernmental Panel on Climate Change. In: Climate Change 2013. Cambridge, UK: Cambridge University Press; 2013

[4] Janssens IA, Freibauer A, Ciais P, Smith P, Nabuurs GJ, Folberth G, Schlamadinger B, Hutjes RWA, Ceulemans R, Schulze ED, Valentini R, Dolman H. Europe's terrestrial biosphere absorbs 7–12% of European anthropogenic CO_2 emissions. Science. 2003;**300**:1538-1542

[5] IPCC. In: I. f. G. E. S. (IGES), editor. Intergovernmental Panel on Climate Change Guidelines for National Greenhouse Gas Inventories. Hayama, Japan; 2006

[6] Stern N. The Economics of Climate Change: The Stern Review. Cambridge and New York: Cambridge University Press; 2007

[7] Wang Y, Li Y, Ye X, Chu Y, Wang X. Profile storage of organic/inorganic carbon in soil: From forest to desert. Science of the Total Environment. 2010;**408**:1925-1931

[8] Guo LB, Gifford RM. Soil carbon stocks and land use change: A meta analysis. Global Change Biology. 2002;**8**:345-360

[9] Lal R, Follett RF, Kimble JM. Achieving soil carbon sequestration in the United States: A challenge to policy makers. Soil Science. 2003;**168**:827-845

[10] Lal R. Challenges and opportunities in soil organic matter research. European Journal of Soil Science. 2009;**60**:158-169

[11] Lal R, Kimble JM, Follet RF, Cole CV. The Potential of U.S. Cropland to Sequester Carbon and Mitigate the Greenhouse Effect. Chelsea, MI: Ann Arbor Press; 1998

[12] Davidson EA, Ackerman IL. Changes in soil carbon inventories following cultivation of previously untilled soils. Biogeochemistry. 1993;**20**:161-193

[13] Post WM, Kwon KC. Soil carbon sequestration and land-use change: Processes and potential. Global Change Biology. 2000;**6**:317-327

[14] Tan Z, Lal R. Carbon sequestration potential estimates with changes in land use and tillage practice in Ohio, USA. Agriculture, Ecosystems and Environment. 2005;**111**:140-152

[15] Reicosky DC. Tillage-induced CO_2 emission from soil. Nutrient Cycling in Agroecosystems. 1997;**49**:273-285

[16] Lal R, Follett RF, Kimble J. Managing U.S. cropland to sequester carbon is soil. Journal of Soil and Water Conservation. 1999;**53**:374-381

[17] Lal R. Offsetting China's CO_2 emissions by soil carbon sequestration. Climatic Change. 2004;**65**:263-275

[18] Lal R, Kimble J. Conservation tillage for carbon sequestration. Nutrient Cycling in Agroecosystems. 1997;**49**:243-253

[19] Murty D, Kirschbaum MUF, McMurtrie RE, McGilvray H. Does conversion of forest to agricultural land change soil carbon and nitrogen? A review of the literature. Global Change Biology. 2002;**8**:105-123

[20] West TO, Marland G. A synthesis of carbon sequestration, carbon emissions, and net carbon flux in agriculture: Comparing tillage practices in the United States. Agriculture, Ecosystems and Environment. 2002;**91**:217-232

[21] VandenBygaart AJ, Gregorich EG, Angers DA. Influence of agricultural management on soil organic carbon: A compendium and assessment of Canadian studies. Canadian Journal of Soil Science. 2003;**83**:363-380

[22] West TO, Post WM. Soil organic carbon sequestration rates by tillage and crop rotation: A global data analysis. Soil Science Society of America Journal. 2002;**66**:1930-1946

[23] Smith P. Carbon sequestration in croplands: The potential in Europe and the global context. European Journal of Agronomy. 2004;**20**:229-236

[24] Lal R. Soil carbon sequestration impacts on global climate change and food security. Science. 2004;**304**:1623-1627

[25] Luo ZK, Wang E, Sun OJ. Soil carbon change and its responses to agricultural practices in Australian agro-ecosystems: A review and synthesis. Geoderma. 2010;**155**:211-223

[26] Rasmussen PE, Allmarans RR, Rhode CR, Roager NC Jr. Crop residue influences on soil carbon and nitrogen in a wheat-fallow system. Soil Science Society of America Journal. 1980;**44**:596-600

[27] Halvorson AD, Reule CA, Follett RF. Nitrogen fertilization effects on soil carbon nitrogen in dryland cropping system. Soil Science Society of America Journal. 1999;**63**:912-917

[28] Halvorson AD, Peterson GA, Reule CA. Tillage systems and crop rotation effects on dryland crop yields and soil carbon in the Central Great Plains. Agronomy Journal. 2002;**94**:1429-1436

[29] Glendining MJ, Powlson DS. The effect of long-term application of inorganic nitrogen fertilizer on soil organic nitrogen. In: Wilson WS, editor. Advances in Soil Organic Matter Research: The Impact on Agriculture and the Environment. Cambridge: Royal Society of Chemistry; 1991. p. 328-338

[30] Alvarez R. A review of nitrogen fertilizer and conservation tillage effects on soil organic carbon storage. Soil Use and Management. 2005;**21**:38-52

[31] Unger PW, McCalla TM. Conservation tillage systems. Advances in Agronomy. 1980;**33**:1-58

[32] Jarecki MK, Lal R. Crop management for soil carbon sequestration. Critical Reviews in Plant Sciences. 2003;**22**:471-502

[33] Franzluebbers AJ, Hons FM, Zuberer DA. Tillage and crop effects on seasonal dynamics of soil CO_2 evolution, water content, temperature, and bulk density. Applied Soil Ecology. 1995;**2**:95-109

[34] Luo YQ, Hui DF, Zhang DQ. Elevated CO_2 stimulates net accumulations of carbon and nitrogen in land ecosystems: A meta-analysis. Ecology. 2006;**87**:53-63

[35] Baker JM, Ochsner TE, Venterea RT, Griffis TJ. Tillage and soil carbon sequestration—What do we really know? Agriculture, Ecosystems and Environment. 2007;**118**:1-5

[36] Blanco-Canqui H, Lal R. No-tillage and soil-profile carbon sequestration: An on-farm assessment. Soil Science Society of America Journal. 2008;**72**:693-701

[37] Paustian K, Parton WJ, Persson J. Modeling soil organic matter in organic-amended and nitrogen-fertilized long-term plots. Soil Science Society of America Journal. 1992;**56**: 476-488

[38] Campbell CA, Zentner RP. Soil organic matter as influenced by crop rotations and fertilization. Soil Science Society of America Journal. 1993;**57**:1034-1040

[39] Alexander RB, Smith RA, Schwarz GE, Boyer EW, Nolan JV, Brakebill JW. Differences in phosphorus and nitrogen delivery to the Gulf of Mexico from the Mississippi river basin. Environmental Science and Technology. 2008;**42**:822-830

[40] Clapp CE, Allmaras RR, Layese MF, Linden DR, Dowdy RH. Soil organic carbon and ^{13}C abundance as related to tillage, crop residue, and nitrogen fertilization under continuous corn management in Minnesota. Soil and Tillage Research. 2000;**55**:127-132

[41] Jagadamma S, Lal R, Hoeft RG, Nafziger ED, Adee EA. Nitrogen fertilization and cropping system effects on soil organic carbon and total N pools under chisel-plow tillage in Illinois. Soil and Tillage Research. 2007;**95**:348-356

[42] Machado S, Rhinhart K, Petrie S. Long-term cropping system effects on carbon sequestration in eastern Oregon. Journal of Environmental Quality. 2006;**35**:1548-1553

[43] Sainju UM, Stevens WB, Caesar-Tonthat T. Soil carbon and crop yields affected by irrigation, tillage, crop rotation, and nitrogen fertilization. Soil Science Society of America Journal. 2014;**78**:936-948

[44] Russell AE, Laird DA, Parkin TB, Mallarino AP. Impact of nitrogen fertilization and cropping system on carbon sequestration in midwestern mollisols. Soil Science Society of America Journal. 2005;**69**:413-422

[45] Alexandratos N. World food and agriculture: outlook for the medium and longer term. Proceedings of the National Academy of Sciences of the USA. 1999;**96**:5908-5914

[46] Liebig MA, Varvel GE, Doran JW, Wienhold BJ. Crop sequence and nitrogen fertilization effects on soil properties in the Western Corn Belt. Soil Science Society of America Journal. 2002;**66**:596-601

[47] Presley DR, Sindelar AJ, Buckley ME, Mengel DB. Long-term nitrogen and tillage effects on soil physical properties under continuous grain Sorghum. Agronomy Journal. 2012;**104**:749-755

[48] Sainju UM, Singh BP, Whitehead WF, Wang S. Carbon supply and storage in tilled and non-tilled soils as influenced by cover crops and nitrogen fertilization. Journal of Environmental Quality. 2006;**35**:1507-1517

[49] Varvel GE. Soil organic carbon changes in diversified rotations of the Western Corn Belt. Soil Science Society of America Journal. 2006;**70**:426-433

[50] Parton WJ, Scurlock JMO, Ojima DS, Gillmanov TG, Scholes RJ, Schimel DS, Kirchner T, Menault JC, Seastelt T, Garcı'a Moya E, Kamnalrut A, Kinyamario JL. Observations and modeling of biomass and soil organic matter dynamics for the grassland biome worldwide. Global Biochemical Cycles. 1993;**7**:785-809

[51] Mudahar MS, Hignett TP. Energy requirements, technology, and resources in the fertilizer sector. In: Helsel ZR, editor. Energy in World Agriculture. New York: Elsevier; 1987. p. 25-61

[52] Cole CV, Flach K, Lee J, Sauerbeck D, Stewart B. Agricultural sources and sinks of carbon. Water, Air, and Soil Pollution. 1993;**70**:111-122

[53] Jarecki MK, Lal R. Soil organic carbon sequestration rates in two long-term no-till experiments in Ohio. Soil Science. 2005;**170**:280-291

[54] Campbell CA, McConley BG, Zentner RP, Selles F, Curtin D. Tillage and crop rotation effects on soil organic C and N in a coarsetextured Typic Haploboroll in southwestern Saskatchewan. Soil and Tillage Research. 1996;**37**:3-14

[55] Angers DA, Bolinder MA, Carter MR, Gregorich EG, Drur CF, Liang BC, Voroney RP, Simard RR, Donald RG, Beyaert RP, Martel J. Impact of tillage practices on carbon and nitrogen storage in cool, humid soils of eastern Canada. Soil and Tillage Research. 1997;**41**:191-201

[56] Yang XM, Wander MM. Tillage effects on soil organic carbon distribution in a silt loam soil in Illinois. Soil and Tillage Research. 1999;**52**:1-9

[57] Cassman KG, Dobermann A, Walters DT, Yang H. Meeting cereal demand while protecting natural resources and improving environmental quality. Annual Review of Environment and Resources. 2003;**28**:315-358

[58] Smith KA, Conen F. Impacts of land management on fluxes of trace greenhouse gases. Soil Use and Management. 2004;**20**:255-263

[59] Li C, Frolking S, Butterbach-Bahl K. Carbon sequestration in arable soils is likely to increase nitrous oxide emissions, offsetting reductions in climate radiative forcing. Climatic Change. 2005;**72**:321-338

[60] Marland G, Fruit K, Sedjo R. Accounting for sequestered carbon: The question of permanence. Environmental Science and Policy. 2001;**4**:259-268

[61] Schlesinger WH. Carbon sequestration in soils. Science. 1999;**284**:2095